The Genesis Column

The Genesis Column

Correlating the Creation Days of Genesis
with the Geologic Column

W. JOSEPH STALLINGS
Foreword by WILLIAM P. PAYNE
Preface by EDWARD N. MARTIN

WIPF & STOCK · Eugene, Oregon

THE GENESIS COLUMN

Correlating the Creation Days of Genesis with the Geologic Column

Copyright © 2018 W. Joseph Stallings. All rights reserved. Except for brief quotations in critical publications or reviews, no part of this book may be reproduced in any manner without prior written permission from the publisher. Write: Permissions, Wipf and Stock Publishers, 199 W. 8th Ave., Suite 3, Eugene, OR 97401.

Wipf & Stock
An Imprint of Wipf and Stock Publishers
199 W. 8th Ave., Suite 3
Eugene, OR 97401

www.wipfandstock.com

PAPERBACK ISBN: 978-1-5326-5554-8
HARDCOVER ISBN: 978-1-5326-5555-5
EBOOK ISBN: 978-1-5326-5556-2

Manufactured in the U.S.A.   10/01/18

Unless otherwise indicated, all scripture quotations in this volume are from the Revised Standard Version (RSV) of THE HOLY BIBLE. Copyright 1946, 1952, 1971 by the Division of Christian Education of the National Council of Churches of Christ in the United States of America. All rights reserved.

To Theadus, my wife,
the greatest of gifts from God and the love of my life

"He who finds a wife finds a good thing,
and obtains favor from the Lord.... She is more precious than jewels."

(Pss 18:22, 31:10b)

To Cassie, our daughter,
whom we love from the greatest depths of our soul,
and who serves Christ with honor

And, to Muffin, Charlie, Gracie, and Rocky,
our beloved canine children,
who have blessed our lives far beyond human words

*These are the generations of
the heavens and the earth when they were created.*

Genesis 2:4

The entire history of the world is preserved in the rocks.

C. Scott Southworth

Contents

List of Diagrams | ix
Foreword | xi
Preface | xv

1 The Earnest Seeking of Truth | 1
2 The Matter of Corruption | 7
3 The Foundations of Correlation | 33
4 The Correlation Model | 52

Bibliography | 145

Diagrams

Figure 1: The Inundative Corruption of *pas ktisis* | 32
Figure 2: Is Correlation Possible? | 51
Figure 3: Existence *Before* Creation Week Correlation | 57
Figure 4: God Day One Correlation | 68
Figure 5: God Day Two Correlation | 72
Figure 6: God Day Three Correlation | 77
Figure 7: God Day Four Correlation | 85
Figure 8: God Day Five Correlation | 91
Figure 9: God Day Six Correlation | 98
Figure 10: God Day Seven Correlation | 104

Foreword

LIKE MANY EVANGELICAL YOUTH of my generation, I quickly learned how to compartmentalize my dispensational faith from my academic pursuits. As far as I was concerned, they were oil and water. I did not try to mix them or to reconcile them. My evangelical faith flowed from my personal experience of God, reading the scriptures, taking part in church, listening to Christian radio, and reading a constant stream of Christian books. On the other hand, my intrigue with science and nature led me into the worldview of naturalism. Because I could not square my religious faith with what I knew to be true in the natural realm, I lost confidence in Christian apologetics.

Since I attended public schools, theories related to biological evolution created a real problem. An image from my 9th grade earth science class still looms large in my mind. I can vividly remember numerous clashes between a Christian student and our science teacher. The student was a Young-Earth creationist. In my opinion, she did not fare well in the discussions. In particular, her arguments for a literal seven-day creation seemed absurd, especially when I considered that the earth did not spin on its axis or revolve around the sun at the beginning of the creation. What was the point of reference for her 24-hour days for which she argued? Also, the evidence for a gradual evolution seemed overwhelming to my young mind. In the end, the Christian student argued that God's Word was true, and we must believe in what it says. What she failed to say was that she equated the truth of the Bible with her fallible interpretation of it.

Two years after I completed college and seminary, while preparing my ordination questions, I realized that I needed to reconcile the Genesis story of creation with my scientifically informed understanding of origins. Since my seminary education did not address this topic, I had to research the issue on my own. To accomplish this, I secluded myself in an out-of-the-way

Foreword

room on an island in Leesburg, Florida. For two weeks, I studied the scriptures and poured over Christian approaches to creation. At the end of my retreat, I declared that I believed in "intelligent design."

Since the universe could not create itself, I concluded that God created mass, time, space, and natural laws. God worked in accordance with them when he created the universe. Furthermore, I argued that the laws and the outcome of God's creative actions could be discovered and observed because creation is the "fingerprint of God." Additionally, I determined that the orderly progress that Genesis records broadly corresponded with the creation sequence that science opined. Unlike any other ancient creation myth, the biblical account did not conflict with the truth that science discovered.

Unfortunately, my science education taught me more than the scientific method. It taught me to think like a naturalist. Because of the philosophy of naturalism, many scientists believe that science and religion cannot be held together. At its core, naturalism affirms that there is a natural cause for everything. For that reason, all knowledge in the universe falls within the pale of natural inquiry via the scientific method. As such, God, angels, ghosts, ancestors, miracles, divine healing, prophecy, revelation, and the supernatural do not exist. When data related to them is examined, that data has to be filtered through the scientific method until a natural cause can be determined. If no natural cause can be determined, the sensory data or the miracle must be rejected.

This truth is illustrated by a lecture I had in a college biology class. The professor (a micro-biologist), discussed the problem of spontaneous generation of life. The accepted theory that life created itself by natural means in pools of water or geo-thermal vents could not be observed in nature or replicated in a lab. To be sure, researchers could produce amino acids and hydroxy acids by means of wetting and drying puddles of water. This enabled polypeptide chains to form. They theorized that these were the building blocks for life. Also, researchers could create artificial cells. Regardless, under no conditions could they create single cell life even though they strongly argued in favor of their theories. Despite this fact, my biology professor still held an unwavering commitment to the scientific theory because of her a priori commitment to naturalism. According to her, in time, we will discover how it happened. When we do, we will replicate it and use it for our own good.

Foreword

When my daughter graduated from high school, my family visited the Smithsonian Institution National Museum of Natural History in DC. After touring the museum and pondering all the exhibits, the final display caught my attention. According to this static display, life came to earth from outer space. It either came on an asteroid or was purposely planted on the earth by aliens. The curator explained that many scientists have looked outside of spontaneous generation to explain the origins of life on earth because they cannot observe the process or reproduce it.

After the presentation, I asked the curator the obvious question. A theory that says life came to earth from another place does not actually solve the problem since it does not show how life created itself. Rather, it pushes it back to another place and time. Then I asked him if God could be the alien that created life on earth. He smiled and told me that he does not mix science with religion.

When scientists delimit God and assume that God does not exist, they impose natural explanation on the data. That is why researchers affirm, "all data is theory driven." No one approaches the data with a blank mind. Instead, the researchers bring their naturalistic commitments with them. This causes them to aver naturalistic causations when the data does not require this. In fact, God is the Creator. He works through the natural process that he created. At critical times, God manipulates those processes to create the outcomes that he desires. Ultimately, Christians affirm that the universe, earth, and humankind are the orderly products of God's purposeful design. Based on this assertion, Christians affirm that God created the creation in order to create us! We are the capstone of his creation and the reason for it.

Ephesians 1:4 says that God chose us in Christ before the foundations of the world. First Peter 1:19–20 refers to Jesus as a perfect lamb without blemish who was chosen to die for us before the creation of the world. Revelation 13:8 refers to the Lamb of God who was slain before the foundations of the earth. What a sobering thought. Paul builds on this theme when he declares that God planned for the glory of humankind before time began. He calls it a mystery that was hidden from the rulers and powers. If they had known about it, they would not have crucified Jesus (1 Cor 2:7–8). In truth, no eye has seen, no ear has heard, and no human mind has conceived the things God has prepared for the redeemed in glory (1 Cor 10:9).

Imagine this. Billions of years ago, before the creation took form, God already was planning for humankind. In fact, he created creation in order to create people! From the beginning, his heart was turned toward humans and his love poured out for humankind. People aren't an accident of evolution. They are the result of a purposeful creation! Knowing this should cause humans to fall before God in praise and thanksgiving. They should trust him and know that he won't abandon them or forsake them. His love for people is eternal and his desire for humankind is sure. For this reason, all human life has purpose and inherent worth.

In conclusion, I wish that I would have been able to have read Dr. Joe Stallings's book, *The Genesis Column*, thirty years ago. While reading it, I discovered that I was an "Old-Earth Progressive Creationist." Not only does Joe show that science and biblical faith are not oil and water, but he actually mixes them together without violating faith *or* science. For sure, he deconstructs naturalism and Young-Earth Creationism. Additionally, his teaching on the positive relationship between the geologic column and the Genesis account will inspire scientifically minded Christians who have always known that they belonged together. On the other hand, his teaching on Adam and Noah will challenge many who hold to the infallibility of Scripture. Even still, his argued conclusions will enable the reader to re-evaluate what Genesis actually teaches about the proto-man and the great flood. Ultimately, we affirm that revelation declares truth, and that science attempts to discover truth. To the extent that science discovers truth, that truth must align with revelation when revelation is properly understood and applied. That is exactly what Joe does for us in this book. Every Christian should read and study this material. In particular, read the footnotes!

William P. Payne, PhD
Professor of Evangelism and World Missions
Ashland Theological Seminary
Ashland, OH

Preface

It is with great pleasure that I offer this preface to Joe Stallings's book, *The Genesis Column*. I must congratulate Joe for writing what appears to be (and what I know to be) a book that has no doubt grown out of all of his great passions in and for matters theological, geological, and hermeneutical (and many other areas of endeavor). Aristotle once wrote that human beings desire to understand. We are understanding beings. We yearn for a great synthesis of the (at least initially, pre-reflective) buzzing confusion of experience, the blooming data of consciousness and reflection, the looming testimony of memory, the integral word of God—the true and the only hope under heaven by which we must be saved—the sometimes alluring pull of reason, and the often centering and hopefully calming voice of one's own tradition. If we have good teachers, we receive a compelling synthesis of these voices. If we have great teachers, we are brought to realize, along with a synthesis of these confluences, that there are many hypotheses, but not all hypotheses are created equal. We learn as well from the great ones in our educational path that while we receive and even fashion compelling syntheses of the data of Word and world, still we are imperfect. Given that we are all time-travelers (just pretty much at a rate that we cannot control), we are still subject to *new data* that will affect our present probabilities and, yes, even our certainties. If the hypotheses we frame and fashion themselves are about God's creation, then get ready: it's going to be a grand synthesis to try to harmonize the data we get from revelation, from the world itself, and from God's work within that world according to the word—and the Word—of God. Our post-reflective interface with each of these data sources must be fairly fine-tuned. And, as we know, this takes a lifelong love and longing for the Triune God to reveal connections, suggest inferences to be made—and some to be avoided, and just plain fill our world with

Preface

insights that stick together to reveal patterns pointing to the one true God behind this greatly designed and beautiful creation.

Joe's book gives us a model to try to harmonize God's word and world along three (at least) common criteria that tend to suggest a theoretical advantage, a stronger theory. For example a good theory must obey the law of non-contradiction—it must be *consistent*. One of the principles of modern logic is that *from a contradiction one can derive anything*. Now that may sound very powerful, but actually it leads to a great self-destructive reality. If one can derive anything—call it x—that means one can also derive not-x. So, both x and not-x are true. But of course, this is the hallmark of error! It would follow that God exists, and that God does not exist. You exist, but you also *don't* exist. The world exists, but there is no world (at any particular time one cares to predicate existence or non-existence, of course). God can do anything in his omnipotence, and it's not the case that God can do anything in his omnipotence. I think you see the point. Once one lets inconsistency in the front, back or side door, one unleashes meaninglessness, and at that point, we all best just eat, drink, and make ourselves merry. Trying to harmonize the Book of Genesis with the best theories found in science, as Joe does within this work, is a tricky business! Many years ago, I served at a school as a dean of philosophy, and a much younger installment of Joe Stallings submitted a similar work then to me, as his doctoral chair, for approval. If he will permit it, here is what I wrote then:

> Joe stays for the most part within the strictures of a pretty strict evangelical offering. Joe does for biological and stratificational reasons hold to an old earth. His approach to try to harmonize science and Christian theology has many ingenious and interesting, new elements. It seems coherent, but there will be those who take umbrage at it because they do not share the presuppositional framework that Joe shares. But these dissonant voices will always be there. Rather, I think, Joe is to be commended for thinking through a complex, multi-variable issue with clarity and genuine truth-seeking humility and charity throughout. It is a pretty impressive attempt, overall, and he is to be commended for his excellent work.

I'm not sure I can summarize the piece as elegantly as I did many years ago. Good work, former self! (But don't get the idea that I—or Joe—buy a B-Theory of Time. More on that inside.)

Secondly, the piece Joe gives us here is *coherent*, as my former self also invoked that word, above. To "cohere" means in the original Latin to "stick together." It's not enough that all of our propositions in our theory together are logically consistent (i.e., no contradiction is formed by believing all the claims of a theory). The relationship between the claims (or propositions) within a theory must give mutual support and a sort of transference of justification and hypothetical relevance and validation one to the other. Again, creating a theory that is coherent is no easy task. Joe gives us a coherent theory, and it's obviously one that he has deeply thought about, prayed about, and talked about for a long time. It is a great testimony to God's majesty that we can hear a person out, allowing them to lay out their theoretical case, and allow our own minds and hearts to "take it in" and consider it—or at least to appreciate it. Joe has earned the right to be heard with his theory. I think he does us a service as fellow Christian believers to try to lay out the biblical revelation concordance model we find within this book and couple that with our own convictions about the two great books, as Galileo called them—the book of biblical revelation and the book of nature. Again, not all will share his presuppositions, but he does an admirable job of explaining, at least, where he came up with his own points of departure and how and why he draws the conclusions that he does. It makes for a very good and thought-provoking read.

Finally, a good theory must be *comprehensive*. The "scope" of the theory must be wide; the theory, as we say nowadays, must have significant *explanatory power*. I think we all know that a good theory of origins must be comprehensive—it must address a wide-ranging set of data about God's creation, human origins, the original state of creation, the origin of evil, suffering, and death, the origin of time, space, matter, and energy, a wide range of other theological, anthropological, philosophical, and geological explananda. I believe Joe has offered a theory that scores well on this criterion, too.

So, what would it be to be given a good theory about origins that meets these three "Cs" of theories? First, the theory would be consistent. Taken any two propositions of the theory, one would have no very fine reasons for thinking that they were inconsistent one with the other. Let's keep in mind here that God is Logic itself; God is a rational being—in fact, God is essentially rational. He is reason itself, and as Augustine famously argued, God is Truth itself. God is the "un-lying" God (Titus 1:2), the one for whom it is 'impossible' to lie (Heb 6:18). This, in principle, it seems fitting, as those made in God's image, to see to be God-like in our thinking—as much as we can—and to be consistent, first of all. By the way, of course it doesn't follow that good theories won't have things that *seem at first blush* to be inconsistent (one thinks of the Trinity and Incarnation here). One might object: but we are imperfect; so why even try to harmonize the present data, if the future is yet to reveal heretofore unknown data yet to be incorporated? The answer is that while theories are the sorts of things that must account for future and yet to be seen data, those data will also be seen *through* theoretical eyes; that is, sometimes it takes a theory to *see data as they are, in all their relevance*, even then to make sense of those data. Examples of this abound. (For instance, what traction would Galileo's theories have gotten without Kepler's theories/laws of planetary motion, or Einstein's special theory without the Michaelson-Morley aether experiments in 1887?) Joe gives us a theory that he has checked carefully for consistency (understanding the provisional nature of nature and the new data it continues to give us before she will, one day, be renewed). We will have to evaluate his data, and render a judgment on this criterion, of course, but by my lights he has run, quite admirably, with due diligence.

As well, a good theory of origins would be coherent—it would, that is, do a good job of showing how data within the theory works together into a mutually supporting framework of interlocking parts and systems that "make sense" together. Finally, the theory must be comprehensive, as we have briefly seen. The theory must have power to explain disparate fields of data, bringing illumination and making sense of distinct fields of inquiry in a way relatively few theories can do. Often comprehensiveness is achieved through a certain elegance of a theory: an attribute which, if true, is both simple yet profoundly powerful in casting light on a whole range of beliefs and convictions that the thinker (and believer) wishes to understand—after all, we all just want (among other things) to *understand*.

What *is*, then, this simple attribute of Joe's theory? I leave that to the reader to explore and find out. For me, though, I believe that Joe's theory is worthy of being judged through the lenses of these tests of consistency, coherency, and comprehensiveness. So, I now invite the reader to put Joe's very fine treatise to the test—or *tests*, as the case might be! Enjoy—it's well worth the ride.

Edward N. Martin, PhD
Co-chair and Professor of Philosophy
Liberty University
Lynchburg, VA

1

The Earnest Seeking of Truth

THE PILGRIMAGE

I HAVE BEEN GREATLY blessed in my life. First, I have been blessed to have been born and raised in a devout Christian family with both of my parents and all four of my grandparents being followers of Christ. I have also been blessed by Theadus, my wife of many years, and our daughter, Cassie, both of whom love Jesus so much as to entrust to him their eternal souls. As for me, although I made a public profession of faith in the summer of 1977, in actuality, I cannot truly remember a time when I did not have a relationship with Jesus. It's beyond me to be able to pinpoint a specific moment when it all began. I suspect my whole life was and is a growing awareness of true reality. Even as an adolescent, I remember marveling at God's world and being fascinated by its intricacies. Everything about the creation was amazing. I knew that the Lord was at work in the tiniest and the greatest processes of nature.

I also remember being very interested in the Bible, although there was much about it that I did not understand. There was (and still is) a certain inscrutable mystery about that very unique book. It was my maternal grandfather (and my namesake), William J. Kiff, who never ceased to astound me with his great knowledge of the Holy Scriptures. He was also one who greatly encouraged me to diligently study the written Word and drink deeply from its truth. Over the years, I have come to love the Bible with an inexplicable passion. Though all parts of the Scriptures draw me in, it seems that I always keep finding my way back to the creation text of Genesis 1–2. I have come to believe that those two chapters tie the whole biblical revelation together. I have also come to believe that those two chapters provide a special intersection with the truth of natural revelation and present a composite picture of the whole of created reality.

Second, I have been blessed with a great interest in science. My granddad encouraged me in this as well.[1] It has been my fascination with God's natural order that has led me to a lifelong fascination with its study. I have been particularly enamored with geology. For me, there has always been something about rocks, strata, and fossils that have unceasingly piqued my curiosity.

As a youngster, I wandered along the roadways, reservoirs, construction sites, and canal embankments of my home neighborhood in Virginia Beach, Virginia, looking for anything of "geological value." More often than not, I carried in the back pocket of my jeans my well-worn copies of the *Fossils* (1962) and the *Rocks and Minerals* (1957) editions of The Golden Nature Guide series. In fact, I still have both of those original copies to this very day.

I also took many trips to and spent many long hours digging away at Rice's Fossil Pit in nearby Hampton, Virginia, looking for that elusive trilobite. It was Mr. William Macon Rice, who saw me there so much and took an apparent liking to me, who explained to me in my frustration that I would not likely find a trilobite in Miocene deposition. However, he also encouraged me to keep digging, assuring me that there were probably the remains of ancient whales still in the pit. I was on top of the world the first time I found a piece of bone left from a four million year old cetacean!

It was these very types of experiences interwoven with my strong Christian upbringing whereby, even as a young boy, I unconsciously developed an integrated view of reality. For me, it was simple. The Bible, of course, was true. It is God's Book. How could anyone doubt it? The world of nature, of course, was true as well. It is God's creation. I constantly observed the world around me, read multitudes of science books about it,

---

1. I even remember Granddad saving his and Grandma's pill bottles for me so that I had extra containers to store rock and fossil samples. Interestingly, in 2011 (he passed in 1987), I ran across an old newspaper clipping that my Mom had saved (from *The Norfolk Ledger-Star*, dated July 30, 1982) in which my granddad wrote a guest article entitled, "Biblical text refutes a creationism-science dispute." Basically, from that article, I discovered that our views of origins essentially mirrored one another. In effect, unbeknownst to me until that time, my granddad—like me—had been a progressive creationist!

and I investigated it every day. It wasn't until I was a religion and philosophy major in college that I became aware that such a view could actually be a matter of controversy. For the first time in my life, I met people who held that a Christian could not rightly believe both that the Bible was true and believe the common geological assertion that the world was very old. Though I thought such a conclusion was bizarre, I began to think critically about the relationship between the Bible and nature, and theology and science. Fortunately, the Lord introduced me to a very significant book that greatly blessed me in my quest.

In 1982, during my second senior year at Atlantic Christian College in Wilson, North Carolina, I first read Bernard Ramm's classic monograph, *The Christian View of Science and Scripture*. Despite my deep struggles with a few of his conclusions, Ramm's concordantist thoughts had a profound confirming influence on my thinking. It was Ramm who gave me "permission" to keep doing what I had been doing—that is, to think with an integrationist perspective. Somehow everything that is true just fits together. There is a certain systematic order to truth and reality. Whenever there seems to be an exception to that idea, it simply means that there must exist some hidden answer that still needs to be found. To this day, I have never stopped believing that postulation. In fact, I suppose that's the very reason this project has come about.

THE BIGGER PICTURE

The study of origins has become a very controversial topic among both the secular and the sacred communities alike. In fact, according to Robert B. Fischer:

> The topics of origins of the universe and the earth, of life, and of humans have probably received more attention than any other one in which information is derivable both from the Bible and from the scientific study of nature. Much of this attention has been engulfed in conflict, often with emotional as well as intellectual overtones. There have been variations in interpretations of what the Bible has to say about these topics, in understandings derived from scientific observations in the realm of nature, and in comparing information obtained from the two sources.[2]

---

2. Fischer, *God Did It, But How?*, 41.

The Genesis Column

This controversy has become especially apparent in the last two hundred years, particularly since the publication of Charles Lyell's *Principles of Geology* in 1830 and later Charles Darwin's *The Origin of Species* in 1859. Today, the debate rages on between creationists and evolutionists, between theistic evolutionists and non-theistic evolutionists, and between Old-Earth creationists and Young-Earth creationists. Yet, despite the seemingly clear and distinctive labels, the lines are not always quite so neatly drawn. This is particularly true concerning the variation of the Old-Earth school that has become known as Progressive Creationism.[3]

Progressive Creationism is possibly understood to be the most controversial position of them all in the current study of origins. On the one hand, other fellow creationists, notably those of the Young-Earth variety, often accuse progressive creationists of being compromisers who actually subjugate the scriptural revelation to natural science,[4] yet cover themselves with a veil of biblicalism in order to make their position more palatable to evangelical Christianity. On the other hand, advocates of evolutionism often make the claim that progressive creationists are actually crafty evangelicals trying to sneak the Bible through the backdoor and into the scientific academy.[5] Either way, Progressive Creationism is often perceived as being somewhere in the middle and thus treated harshly from multiple angles within the classic origins discussion.

In reality, a thorough study of Progressive Creationism shows that its intent is to be neither compromising nor subversive, but rather to be an honest evidential postulation based on an earnest seeking of truth. Progressive Creationism[6] is simply a genuine attempt to reconcile the truth of Holy Scripture and the truth of nature as related to the origination and perpetuation of the created order. It holds to the fundamental precept that the scriptural revelation of God (2 Tim 3:16), as properly understood using

---

3. It is very possible that Russell L. Mixter (Wheaton College) coined the term "progressive creation" in the late 1930's. His ideology, however, seems to be leaning more toward theistic evolution rather than toward a true form of creationism. See Numbers, *The Creationists*, 195–204.

4 Batten and Sarfati, *15 Reasons*, 13. See also, Chaffey and Lisle, *Old-Earth Creationism on Trial*, 31–36.

5. Scott, *Evolution Vs. Creationism*, 62–63.

6. The word *progressive* as used here is not to be confused in any way with the currently popular term describing the more liberal-leaning branch of Christian theology. The term *Progressive Creationism* simply refers to the understanding that God created the universe in successive actions set within *epochal* "Day-to-Day" (progressive) steps with a final and ongoing anthropic completion in mind.

the theological disciplines, and the natural revelation of God (Ps 19:1–4; Rom 1:20), as properly understood using the scientific disciplines, must show some form of basic concordance. Both of God's revelatory "books," the Book of Scripture and the Book of Nature, are thus considered to be evidence, of which, when properly interpreted together, reveals the comprehensive and ultimate truth of existence.

Although Progressive Creationism in its modern form has existed at least since the aforementioned Bernard Ramm (1916–1992) in the 1950s,[7] its most prominent current advocate arguably is Hugh Norman Ross (b. 1945).[8] While there are other current influential scholars who are progressive creationists, including Robert C. Newman,[9] David Snoke,[10] and C. John Collins,[11] the claim of Rossian prominence is supported by his extremely prolific publication record.[12] It is also supported by the fact that his work remains the top target of attacks from both evolutionists and Young-Earth creationists alike.

This text is written from the same general viewpoint as that of Ross, but with some significant modifications. Without apology and with much aplomb, I too am a devout Old-Earth progressive creationist. Over the years, I have been greatly and very positively influenced by the works of both Ramm and Ross. (They, in particular, will be referenced frequently in this book.) However, being a critical scholar, I have also long seen the need to add another dimension to Progressive Creationism in order to enhance and reinforce its systematic strength. Therefore, in this volume,

---

7. See Ramm, *The Christian View of Science and Scripture*. This book was originally published in 1954. It should also be noted that there have been other prominent contemporary evangelical scholars who have held to an "old earth" view, including B. B. Warfield, John Walvoord, Francis Schaeffer, and Gleason Archer Jr. Further, a number of past scholars have held to a version of Day-Age Creationism, which is compatible to Progressive Creationism. We mention Ramm specifically because of the major impact of his classic book.

8. Ramm was a professional theologian who also studied science; Ross is a professional scientist who also studies theology. Both scholars are long known to be seekers of truth who hold to a wide integrationist understanding of reality.

9. Newman and Eckelmann Jr., *Genesis One*.

10. Snoke, *Biblical Case for an Old Earth*.

11. Collins, *Genesis 1–4*; also, *Science & Faith*.

12. Unofficially, as of 2018, Ross has authored at least sixteen major books, contributed to four multi-author volumes, and published many articles on creationism and Christian apologetics. He is also a speaker-lecturer in great demand throughout the United States and the world. For more, see his *Reasons to Believe* website: www.reasons.org.

I will present the *Genesis Column Model*, which is a correlation model of the scriptural Creation Days of Genesis and the geologic column of nature.

May the Kingdom of God be advanced, may God's Holy Church be blessed, and may God himself be glorified.

2

The Matter of Corruption

NOT EVERYTHING HELD TO be true by mainstream Old-Earth Creationism is properly compatible with an evangelical understanding of Scripture. One such matter in particular must immediately be addressed—even before we begin presenting the correlation model; namely, the idea as to whether or not death and corruption existed and inundated the natural order long before the fall of Adam. According to the classic Old-Earth understanding,[1] death and corruption has its origin in the creative will of God.[2] In fact, it is an understanding that seems to be required by the presence of epochal time before the creation of humanity. After all, the geologic record as understood by modern science is replete with evidences of death and corruption (i.e., fossils, etc.) millions of years before the first appearance of man on Earth.[3] How is it possible, then, for a faithful Christian, with both a strong

---

1. For example, see Stoner, *A New Look at an Old Earth*, 42, 50–53.

2. See Miller, "And God Saw That It Was Good," 85–93. He explains: "Creation is good, and the death and pain embedded within it are part of God's will and purpose for it. Creation is not a fallen thing" (92). For further, also see the following: Schicatano, *The Theory of Creation*, 144–54; Neyman, "Death Before the Fall of Man"; and Phillips, "Did Animals Die before the Fall?," 146–7.

3. Rudwick, *Worlds before Adam*, 11–552. In this work, Rudwick thoroughly substantiates the common geohistorical premise that long before the appearance of humanity many faunas continuously appeared and vanished over the course of deep time. For example, he briefly outlines (22–23) the observational contributions and postulations of early geologist Georges Culvier to the foundations of modern stratigraphic geohistory. According to Rudwick, "Cuvier presented a forceful and eloquent case for treating geology as a *historical* science, and for hitching the short span of recorded human history

evangelical theology (including a high view of Scripture) and an adherence to orthodox geohistory, to be able to make sense of such a chronology?

The Original Creation

In order to answer, it is appropriate to examine the text of Genesis 1 in order to determine what it communicates on its face about the *qualitative state* of the final and completed form of the original created order. Here is a systematic run-down of the text:

- *Creation Day One* (Gen 1:3–5): God saw that the light was "good" (Hebrew *tob*);
- *Creation Day Two* (Gen 1:6–8): God separated the firmament, but made no qualitative pronouncement;
- *Creation Day Three* (Gen 1:9–13): God saw that the land and sea was "good" (Hebrew *tob*); God saw that the land vegetation was "good" (Hebrew *tob*);
- *Creation Day Four* (Gen 1:14–19): God saw that the sun, the moon, and the stars were "good" (Hebrew *tob*);
- *Creation Day Five* (Gen 1:20–23): God saw that the swimming and flying *nephesh* creatures were "good" (Hebrew *tob*); and,
- *Creation Day Six* (Gen 1:24–31): God saw that the *nephesh* land creatures were "good" (Hebrew *tob*); God then created humanity, both male and female, in his own "image" (Hebrew *besalmenu*) and "likeness" (Hebrew *kidmutenu*) and "blessed" (Hebrew *wayebarek*) them; God then saw everything that he had made and proclaimed that it was "very good" (Hebrew *tob meod*).[4]

---

on the tail end of an inconceivably longer span of prehuman history" (22). See also, Rudwick, *Bursting the Limits of Time*, 1–11. Here he states that "a focus on the magnitude of the timescale [i.e., akin to John McPhee's *deep time* concept] has masked the much greater significance of the 'deep history,' as I shall call it, that fills up the vast tracks of deep time. Above all, it obscures an even more dramatic feature, namely the change from regarding human history as almost coextensive with cosmic history, to treating it as just the most recent phase in a far longer and highly eventful story, almost all of it *prehuman*" (2).

4. See Sailhamer, "Genesis," 24–38.

The Matter of Corruption

It is clear that God understood the individual parts of his original creation to be "good," and then the totality of that creation (culminating in his *blessed image-bearer*, humanity) to be no less than "very good." This is the definitive assertion of the Genesis 1 text. Note that the difference in the creation being proclaimed *good* (Hebrew *tob*) and being proclaimed *very good* (Hebrew *tob meod*) is the additional divine lavishing of creation with its Crown (humanity), and then with the further specific lavishing of the Crown of creation with the multiple adornments of divine image (Hebrew *besalmenu*), divine likeness (Hebrew *kidmutenu*), and divine blessing (Hebrew *wayebarek*). Before the created order was finally complete, God added the personal tri-fold touch of himself. The final package included, in effect, God's own *very good* self-reflection (cf. Pss 86:5; 100:5; 106:1; etc.).

The Great Question

In light of *all creation* ("everything," Gen 1:31) being considered as "good" and finally, as "very good" by divine proclamation, Gary Emberger presents this intriguing thought:

> Suffering, extreme pain, and death are part of the natural world. Consequently, Christians tend to view the natural world with ambivalence. We believe God created it, pronounced it good, and rules over it. We read that God provides food for the lion and the raven (Job 38:39, 41), birds of the air (Matthew 6:26), notes the death of sparrows (Luke 12:6), sports with leviathan (Psalm 104:26), creates the beauty of the flower (Luke 12:27), and that all of creation praises him (Psalm 148). The complexity, beauty, and apparent design of our world is presented in the Bible as a clear witness to God's "invisible qualities—his eternal power and divine nature" (Romans 1:20). And yet, what about aging and death? What about disease, parasites, predators, droughts, earthquakes, birth defects, floods, blindness, mental retardation, and accidents? Are these the stuff of God's good creation?[5]

In Genesis 1, God proclaimed six times in succession that his creation was "good" (Gen 1:4, 10, 12, 18, 21, 25). After the creation of humanity, he then proclaimed that "everything" was "very good" (Gen 1:31). Should the things named by Emberger—death, aging, parasites, earthquakes, etc.—be accounted as being *very good* by divine proclamation? Or, has something

---

5. Emberger, "Theological and Scientific Explanations," 150–58.

devastating happened to the *very good* creation since God made his qualitative announcement? As Gerald Wheeler clarifies:

> The conservative Christian believes that God created the universe and its basic life forms. According to Scripture, when God originally made life, He considered it "good" (Genesis 1:25). Did God judge the goodness of His creation by a different standard, or has something happened to it in the meantime?[6]

This is a very important question for Christians. It brings to bear the problem of evil.

The Great Problem

The issue of the problem of evil in the world is known as *theodicy* (Greek *theosdike*; lit., "God justice"). William H. Willimon defines theodicy as "the philosophical attempt to justify the ways of God to humanity, an attempt to think about what God does with evil and why."[7] Robert M. Adams calls it "a defense of the justice or goodness of God in the face of doubts or objections arising from the phenomena of evil in the world ('evil' refers here to bad states of affairs of any sort)."[8] Perhaps the great problem of theodicy is summed up best by John Hick: "If God is perfectly good, he must want to abolish all evil; if he is unlimitedly powerful, he must be able to abolish all evil: but evil exists; therefore either God is not perfectly good or he is not unlimitedly powerful."[9] This is the great dilemma faced by Christians, especially as related to apologetics and evangelism. How can an omnipotent, omniscient, omnipresent God, who is love, create and perpetuate a world and universe that is filled with evil and death? This very issue is often the greatest stumbling block for non-Christians in receiving and accepting the biblical reality of the saving Gospel of Jesus Christ. As Emberger says, "We find it difficult to reconcile ourselves to the presence of evil in a world created by an omnipotent God of love."[10]

---

6. Wheeler, "Cruelty of Nature," 32–41.
7. Willimon, *Sighing for Eden*, 34.
8. Adams, "Theodicy," 910.
9. Hick, *Evil and the God of Love*, 5.
10. Emberger, "Theological and Scientific Explanations," 150.

The Two Great Theological Explanations

Many efforts at explaining the presence of evil attempt to neatly separate moral evil (i.e., evil originated by humanity: murder, greed, lying, etc.) from natural evil (i.e., evil originated apart from humanity: hurricanes, extreme pain in animals, tetanus, etc.). However, these neat divisions are usually found to be overtly artificial. For many scholars, there is much uncertainty as to the actual cause of evil and as to what should actually be considered evil.[11] Other efforts, such as the Alfred North Whitehead explanation, who following Hick's line of thought, have attempted to deal with evil by positing an essentially good, but non-omnipotent God—namely, a Deity who would like to dispose of all evil, but who does not have the sufficient power to do so.[12] Still other models focus on final outcomes rather than on evil origination; in the end, God will turn out to be very good in that the evils of this life will pale in comparison to the blessings of the next.[13] Yet, despite the questions and the uncertainties and the multitudes of attempts to understand, classic theology is able to clearly present two major antithetical views of evil: the Augustinian explanation and the Irenaean explanation.

The Augustinian Explanation

Augustine (Bishop of Hippo, 354–430 CE) understood evil to be the result of the creaturely (human or angelic) misuse of freedom. Rational beings, gifted with free will from God, made wrong sinful choices. These choices resulted in the corruption of God's perfect ("good"/"very good") creation.

In this view, both the moral evil and natural evil of the created universe and world are truly evil and had their origin at the Adamic fall in the Garden of Eden. This "true evil" is often understood to be God's subsequent judgment upon humanity and upon the world in which humanity had been placed as overseers. Emberger explains:

> Man and at least the higher animals were created immortal, not susceptible to illness or aging. Carnivores did not exist; animals were herbivores. Evolution—or at least macroevolution—was not one of God's creative mechanisms. The Second Law of Thermodynamics—stating that all things move toward increasing

11. Ibid., 151.
12. Adams, "Theodicy," 910.
13. Ibid., 910–11.

> disorder—either was not in effect or was neutralized by God's continual sustaining or organizing power. Into this world came moral sin, the willful turning from God, with natural evils quickly following. All of creation was corrupted. Parasites, predators, and disease organisms are postulated to be post-Fall microevolutionary developments. Aging and death began. Sin was followed by further judgment with the Flood and accompanying movements of the earth's crust giving rise to earthquakes, volcanoes, the ice age, massive extinctions (as evidenced by fossils), and other natural evils. In short, natural evil exists because of man's sin.[14]

This particular version of Augustinian theodicy is often associated with Young-Earth Creationism. However, some *Old-Earth* Christians, those who are not comfortable with the idea that God orchestrated natural evil into the original creation, posit that the fall of Lucifer was the cause of the most ancient natural evil long before the creation and fall of Adam. Such thinking typically goes like this:

> Satan, cast down to earth with the other fallen angels, attacked God's perfect creation, disrupting and distorting it, causing earthquakes, volcanoes, disease, predation, and death. The fall of Adam and Eve resulted in further evils including human death and the further corruption of the natural world due to man's broken relationship with God and thus his broken relationship with God's creation. This theodicy leaves room for evolutionary processes but it is an open question as to how much "good death" could have occurred.[15]

All variations of the Augustinian explanation share the common assertion of the "innocence" of God and the responsibility of his creatures for the advent of natural evil into the created order.[16]

The Irenaean Explanation

Irenaeus (Bishop of Lyons, c. 125–202 CE) believed that the present world, with its full content of natural evil, is essentially the world that God originally created. The divine intention was to provide an environment of pain

---

14. Emberger, "Theological and Scientific Explanations," 3.
15. Ibid. Note: This is often formatted into some variation of the Gap Theory. For example, see Hagee, *Revelation of Truth*, 9–24. Also, see Johnson, *Bible, Genesis & Geology*.
16. Emberger, "Theological and Scientific Explanations," 3.

and struggle whereby people could develop through trial toward perfection and ultimately choose to fulfill God's good purpose for the human race. According to Emberger:

> Here, ultimate responsibility for natural evil in the world is placed on God. Man, arising through an evolutionary process, imperfect and immature, came to a point in his development where he was capable of fellowship with God, capable of acknowledging his presence. But God could not "force" the next phase of his purpose for man—to bring into existence children of God, beings who will freely choose to love God and to grow in his knowledge. In order to accomplish this second phase it was necessary to place humans in an ambiguous world much like our current world . . . [I]f God were unambiguously present in the world, man would not truly be free to choose to come to God, he would be overwhelmed by God. And so, God created an ambiguous world—a world with pointers to himself, but also a world where he could be seen as absent . . . It was a world filled with good things but also real evils and real challenges . . . Only in this kind of world, through a long process of creaturely experiences, both good and bad, could people freely choose to love God and develop the kind of goodness God values.[17]

In the Irenaean view, the fall, though not automatic, was virtually an expectation. The human race was struggling in a hostile environment where God did not seem to exist. Thus, humanity chose to deny God in his ambiguity and live in a manner as if the natural world was the extent of reality. Largely due to God's decision to remain somewhat distant, immature humanity chose to become self-centered, self-exalting, and worldly. This is the essence of the Adamic rebellion against God.[18]

The Quick Comparison

These are the two primary antithetical explanations for theodicy. The Augustinian explanation, which has been the dominant view—particularly in Western Christianity, posits a perfect original creation that became marred by the bad use of free will by God's rational creatures. With this view, the creatures are ultimately responsible for the evil in the universe. Contrarily, the Irenaean explanation posits a very hostile original creation that was

17. Ibid., 3–4.
18. Ibid., 4.

filled with natural evil from its inception whereby God's rational creatures were, in effect, almost expected to choose wrongly. With this view, God is ultimately responsible for the evil in the universe.

THE SUGGESTED ALTERNATIVE

The Modified Augustinian Explanation

We propose a modified Augustinian explanation. It is a classic Augustinian theodicy in that it holds to the traditional understanding that Adamic sin indeed caused death and corruption to enter and affect the entire (originally pristine) created order. It is modified in that it will be applied to an Old-Earth paradigm (as opposed to a model of recent creation) without focusing on the fall of Lucifer and/or the Gap Theory.

The Meaning of Good

Having established previously that the scriptural revelation most clearly asserts that the original creation was "good," it is necessary to discuss just what God meant when he announced it as being "good." It is important to note, first, that the initial condition of the primordial Earth is that of "without form [having no form] and void [empty]" (Hebrew *tohu wabohu*). It was also covered in "darkness" [without light] (Hebrew *choshek*). The details of the picture that is conveyed are conditions of proto-planet chaos and bleak uninhabitability; they are conditions quite untenable for human existence (or for the existence of any life whatsoever). At that primitive point of natural history, the Earth existed, but in a very early "not-yet" state (that is, not yet what it was intended by God to eventually become, i.e., the home of humanity). As John H. Sailhamer says:

> The general context of chapter 1 suggests that the author meant the terms "formless and empty" to describe the condition of the land *before* God made it "good." Before God began his work, the land was "formless" (*tohu*), and God then made it "good" (*tob*). Thus the expression "formless and empty" ultimately refers to the condition of the land in its "not-yet" state—the state it was in before God made it "good." In this sense the description of the

land in 1:2 is similar to that in 2:5–6. Both describe the land as "not-yet" what it shall be.[19]

This implies that there is an important *contrast* developing in the text. It is, in effect, a contrast between that which is *not* (yet) *good* and that which is (becoming) *good*. There is also an important *context* developing as well. It is the context of a human-friendly perspective. Sailhamer correctly states that "the remainder of the [creation] account is a portrayal of God's preparing the land for man."[20] This is very important for, in answer to Wheeler's earlier question, it suggests that God's assessment of his creation as "good" would be an assessment that humans would probably understand and agree with and not "a different standard of goodness." After all, God was making the world for people. It seems that if Adam, either in his pristine or fallen state, were present with God as he worked through the Creation Days, he would be nodding in the affirmative along with God (as well as marveling in amazement at God) with the fine completion of each new thing. This image defies the thought that the original creation would include those things that humanity distains as being natural evil. In other words, the tenor of the Genesis 1 text points to the idea that God's conception of original "good" is not a different standard than that of divine perfection in the classic human denotation (cf. Jas 1:17 = parallel use of God's "good" and God's "perfect"[21]).

According to Cuthbert A. Simpson, God's pronouncement of "good" means "that it [creation] was the perfect reflection of his thought."[22] While it is indeed true that God's thoughts are not our thoughts (Isa 55:8–9), it is most plausible to assert that the divine thought at original creation would not be reflective of death, decay, and corruption (i.e., those things that rational humanity bearing the divine image would consider to be evil). It is, quite frankly, a stretch to consider otherwise. This postulation is further substantiated by the idea that the ("not good") formless and void stage of Earth that is conveyed in Genesis 1:2 was in the process of being made "good" (i.e., taking form and becoming abundantly full) over the course of God's six Creation Days.

Throughout the Creation Week, God was making the Earth into a "good" environment for people to live. The "not good" conditions of

---

19. Sailhamer, "Genesis," 24.

20. Ibid.

21. Mounce, *Expository Dictionary*, "Good," 300—Greek *agathos* (G18) = "good"; "Perfect," 506—Greek *teleios* (G5046) = "perfect."

22. Simpson, "Genesis" [Exposition], 469.

Genesis 1:2 (namely, those that human beings would most assuredly understand as being "not good") would be transformed by God into "good" conditions (namely, those that human beings would most assuredly understand as being "good") via the Creation Week. This interpretation appropriately incorporates the contrast (not good / good) with the context (a developing human-friendly perspective) that is inherently present in the text of Genesis 1.

Those (like Ross) are wrong to claim that "sin causes humans to react negatively" to the presence of death and corruption.[23] Such thinking is nonsensical. Certainly God is able to use the current existence of death and corruption—really, any and all things—for the ultimate fulfillment of his good purposes (see Prov 16:4; Rom 8:28). Yet, in so doing, he crafts, molds, and works through otherwise tragic circumstances and events to accomplish his ultimate intentions.

Make no mistake—these things that humanity knows as death and corruption are indeed tragic and evil. They are not good and never have been; they were not a part of the original creation. Keith B. Miller makes the very Irenaean claim that the original creation has a "cruciform character"[24] and that this condition is a "God-given character."[25] This is a "cart before the horse" absurdity. It is inconceivable as to why God would cause the original creation—*before* Adamic sin—to have a *cruciform* character.[26] Alternatively, Schaeffer provides an Augustinian view that is much more plausibly confluent with the Holy Scriptures. Schaeffer describes the universe following the Adamic fall as becoming "the abnormal universe."[27] He says, "In other words, at this point [the fall] the external world is changed."[28] In contrast

---

23. Ross, *Matter Of Days*, 109. If anything, it seems that sin might cause humans to react positively to it.

24. Miller, "God Saw That It Was Good," 85.

25. Ibid., 92.

26. The image of the cross is associated with atonement and sacrifice. Is the application of this image intended to be a pre-Calvary preliminary (even anticipatory) semi-atonement of some sort whereby all creation—including people—are purposefully subjected to suffering and corruption *before* the actuality of Adamic sin? Though the Lamb of God was slain before the foundation of the world within the divine Mind, the efficacy was not made manifest in space-time until the *post-fall* Christ Event. Thus, a cruciform original creation interpretation is not only a misappropriation of basic Christian theology (placement of an unnecessary "pre-cross" in the mix), but potentially blasphemous as well (Christ's sacrifice insufficient?).

27. Schaeffer, *Genesis in Space and Time*, 94.

28. Ibid., 95.

The Matter of Corruption

to Ross, his idea is that the fall resulted in the external, physical universe changing from its original normal condition to its current abnormal condition (perhaps like cancer appearing in a previously healthy human body). Shaeffer elaborates:

> Christianity as a system does not begin with Christ as Savior, but with the infinite-personal God who created the world in the beginning and who made man significant in the flow of history. And man's significant act of revolt has made the world abnormal. Thus there is not a total unbroken continuity back to the way the world originally was. Non-Christian philosophers almost universally agree in seeing everything as normal, assuming things are as they always have been. And, of course, this is very important to the explanation of evil in the world. But it is not only that. It is one way to understand the distinction between the naturalistic, non-Christian answers (whether spoken in philosophic, scientific or even religious language) and the Christian answer. The distinction is that as I look about me I know I live in an abnormal world.[29]

The depth of Schaeffer's powerful assertion should not be missed. He is claiming that the fall explains the infusion of all evil (both natural and moral) into the created order. Any other explanation, whether couched in philosophic, scientific, or religious words (as does Ross and others) is a *naturalistic* and *non-Christian* explanation.

The Christian answer is the better answer. Though death and corruption came into this world as a result of Adam's revolt, reality must be viewed in the context of divine redemption and renewal for people and the entire physical order. Due to the work of God in Christ, evil (with all of its degradative manifestations: death and corruption)—a foreign invader of God's original perfect creation—does not have the final word. Through Christ, God has provided a reversal. Not only will all forms of death be divinely cast away and destroyed (Rom 5, esp. vv. 18–21;[30] Rev 20:14[31]), but

---

29. Ibid., 97.

30. Rom 5:18–21, "Then as one man's trespass led to condemnation for all men, so one man's act of righteousness leads to acquittal and life for all men. For as by one man's disobedience many were made sinners, so by one man's obedience many will be made righteous. Law came in, to increase the trespass; but where sin increased, grace abounded all the more, so that, as sin reigned in death, grace also might reign through righteousness to eternal life through Jesus Christ our Lord."

31. Rev 20:14, "Then Death and Hades were thrown into the lake of fire."

all forms of corruption will "pass away" (Rev 21:1–4)[32] through the power of the Living God. "And he who sat upon the throne said, 'Behold, I make *all things* (Greek *pas*) new'" (Rev 21:5).[33] The coming New Creation will be completely pervasive.

The Role of God and the Effects of the Fall on Nature

We present four key assertions. First, God created the universe in an original condition of true good ("very good"), namely, without evil—natural or otherwise. Second, humanity—under the encouragement of Satan—made a rebellious and sinful decision in defiance of God. This decision was the catalyst that brought evil into the previously perfect creation marring it with death and corruption. Third, there was a point in the chronological past (2 ka—c. 33 AD) when God acted decisively through the Christ Event for the redemption of humanity and all of which humanity holds dominion. Fourth, there will be a point in the chronological future when God will bring about an eschatological conclusion to this fallen order and usher in a New Creation (one that will be completely and eternally untainted by any and all brands of evil and its corollary effects).[34]

With these assertions as the backdrop, it is appropriate to focus on the Adamic fall and the response of God. Initially, God did not immediately act in chronological time to instigate the Christ Event or the New Creation. In the omniscient sensibilities of the divine Mind, the fullness of time (in *kairos*) had not yet arrived for either event (in *chronos*). Rather, God quickly sought to keep the damage under some semblance of control. The first step was to confront the offenders with the magnitude of the sin so

---

32. Rev 21:1–4, "Then I saw a new heaven and a new earth; for the first heaven and the first earth had passed away, and the sea was no more. And I saw the holy city, new Jerusalem, coming down out of heaven from God, prepared as a bride adorned for her husband; and I heard a loud voice from the throne saying, 'Behold, the dwelling of God is with men. He will dwell with them, and they shall be his people, and God himself will be with them; he will wipe away every tear from their eyes, and death shall be no more, neither shall there be mourning nor crying nor pain any more, for the former things have passed away.'"

33. All Scripture references are taken from the Revised Standard Version (RSV) of *The Holy Bible* unless otherwise indicated.

34. See Wesley, "General Deliverance," 437–50. He presents the comprehensive biblical panorama of Original Creation to Present Creation to New Creation.

that ultimate reconciliation between humanity and God could be eventually accomplished.[35] As William A. Dembski states:

> In the fall humans rebelled against God and thereby invited evil into the world. The challenge God faces in controlling the damage resulting from this original sin is to make humans realize the full extent of their sin so that, in the fullness of time, we can fully embrace the redemption in Christ. Only in this way can we experience full release.[36]

In order to provide some clarity in understanding the divine response to the Adamic fall, it is important to examine the classic patristic teaching concerning the will of God. According to this teaching, the will of God has three basic forms. First, God can utilize his *active will*. This means that God brings about a desired end by direct action (e.g., God calls a person to a task). Second, God can utilize his *providential will*. This means that God causes the created order to have certain consistent characteristics inherent by particular design (e.g., God institutes the laws and processes of nature). Third, God can utilize his *permissive will*. This means that God takes a short step back and allows a bit of wiggle room so that certain free agents or situations can simply run their course reasonably unhindered (e.g., God permits Satan to act in a specific manner).[37]

According to Dembski, all three variations of divine will were expressed in "the disordering of creation via natural evil" (i.e., the results of the fall).[38] He says:

> Genesis 3:17–18 suggests that God actively wills thorns and thistles (which symbolize the material effects of the Fall). Once

---

35. Note the order and magnitude of divine confrontation. First, God confronts Lucifer: "Because you have done this, cursed are *you*" (Gen 3:14). Second, God confronts Eve: "I will greatly multiply your pain in childbearing" (Gen 3:16). Third, God confronts Adam: "Because you have listened to the voice of your wife, and have eaten of the tree . . . cursed is the *ground* because of you . . . in toil you shall eat of it all the days of your life" (Gen 3:17). In successive order, God curses Satan to eternal corruption and death, yet God disciplines Eve (with increased labor) and Adam (with increased labor) for future restoration; there is no divine curse placed upon humanity. Note also that the original created order (and by extension, the present version) is cursed to corruption and death as well—yet even the creation itself will ultimately be renewed at the eschatological consummation (Rev 21:1). Only Satan—the root cause of the evil—is left with a destiny without the hope of redemption (Rev 20:10).

36. Dembski, *End of Christianity*, 145.
37. Ibid., 145–46.
38. Ibid., 146.

> thorns and thistles are there, however, they perpetuate themselves through natural processes of reproduction. This suggests that God is also providentially willing thorns and thistles. We might say, therefore, that the arrival of thorns and thistles reflects God's active will, their survival reflects his providential will. In addition, nature consequent to the Fall exhibits a nastiness and perversity that seem hard to attribute to the active will of a loving God. Vipers, viruses, and vermin seem more appropriately attributed to God's permissive will, the permission going to Satan.[39]

Notice what Dembski is asserting. The Adamic fall was the occurrence of "the disordering of creation via natural evil." This implies that there was once, prior to the disordering, a state of order in creation. It means, in Dembski's understanding (and we concur with that understanding), that the original creation was changed by the entrance of natural evil at the fall and the result was a disordering of a previously orderly existence: from good to evil. It also means that God's will was involved as well in various degrees: actively, providentially, and permissively.

This fits well with what the Apostle writes in Romans 8:19–21, "For the creation waits with eager longing for the revealing of the sons of God; for the creation was subjected to futility, not of its own will but *by the will of him who subjected it in hope*; because the creation itself will be set free from its bondage to decay and obtain the glorious liberty of the children of God." It is important to understand the workings of divine will in the context of ultimate redemption and final restoration.[40] In his omniscient foreknowledge, God had a plan in place "even before the foundation of the world" (Rev 13:8; cf. 1 Pet 1:18–21) to bring about a reversal of this "bondage" to "futility" and "decay." Due to the Adamic fall, both humanity and nature became linked in bondage to degradation in accordance with the will of God (whether it is attributed to his active will, providential will, permissive will, or a combination or totality of his wills). Dembski offers this thought:

> I submit that the link between human sin and natural evil is far from arbitrary . . . God is fully justified in linking the two. The broad principle that justifies linking human sin and natural evil is

---

39. Ibid.

40. See Harrison, "Romans," 94. Harrison states, "The one who subjected the creation is not named. Some early Fathers assumed that Adam is in view. Others (e.g., Godet) incline to the notion that Satan is meant. But by far the most natural interpretation is that which postulates God as the one who did the subjecting." We certainly concur with Harrison.

humanity's covenant headship in creation. Humans are the priests of creation, offering a world given by God back to God. God, having placed humanity in this position, holds creation accountable for what its covenant head does. God's dealings with creation therefore parallel his dealings with humanity. If God's relation with the covenant head goes awry, so does his relation with all that the covenant head represents (in this case, the world).[41]

Just as God, prior to the fall, subjected humanity and the natural order to abundance and immortality, he likewise subjected humanity and the natural order to futility and mortality after the fall. Yet, it should be remembered that the sinful choice of humanity, the covenant head of creation, instigated the act of divine will.

The Drastic Auto-Effect of the Fall

We believe that a great weakness of mainstream Old-Earth Creationism is its advocacy of death and corruption *prior* to the Adamic fall by the design and ordering of God. Such an advocacy is determined to be unsatisfactory and quite problematic within the framework of classic orthodox evangelical understanding.[42] This becomes readily apparent with an appropriate theological-philosophical approach to the normative "effects of the fall" text found in Romans 8:22–23. Paul wrote these important words: "We know that the whole creation has been groaning in travail together until now; and not only the creation, but we ourselves, who have the first fruits of the Spirit, groan inwardly as we wait for adoption as sons, the redemption of our bodies."

The Two Important Points

There are two important points to make as to the message of Romans 8:22–23. First, the text makes an unmistakable connection between the physical creation and the physical bodies of human beings. In fact, the point is that both have been exposed to corruption and are, as a consequence, "groaning in travail together until now" awaiting a redemption (i.e., a restoration or a *new genesis*; cf. Matt 19:28—Jesus specifically uses the Greek *palingenesia*

---

41. Dembski, "End of Christianity," 147.
42. Henry Morris III, "Being Like Him."

["new beginning"] in reference to the coming "new world"[43]) sometime in the chronological future. This flies in the face of Ross's belief that the fall only affected the human spiritual dimension and had little or no direct affect upon the natural order. Notice the power and the direction of the words used in the text: "groaning in travail" (i.e., "not good") awaiting "redemption" (i.e., "good"). This thought seems to parallel the imagery of the creation text of Genesis 1 as God makes the "formless and void" proto-Earth of Genesis 1:2 into the "very good" terrestrial home for humanity in Genesis 1:31. Note also that the descriptive language used by Paul in referring to the current state of creation ("groaning in travail") is reminiscent of labor pains. In context, it is certainly not hard to be reminded of the divine discipline (increased labor pains) put upon Adam and Eve after the fall (Gen 3). However, the thought is that the labor pain, though long and intense, is actually part of the divine (post-fall) process of remaking the degenerated "not good" into the regenerated "good" and will ultimately issue in a rebirth of new life and new abundance. Cyril E. Blackman puts it simply, "The strong and colorful verbs [of Rom 8:22–23] express the duress of life apart from Christ."[44] Blackman's succinct thought is quite significant, for it substantiates that Paul's "strong and colorful" image of creation's futility ("groaning," etc.) implies the reason for its condition ("duress"): relational separation from God (i.e., due to the Adamic fall). Therefore, it is most plausible to posit that Romans 8:22–23 does indeed teach that natural evil invaded the entire created order—both the organic and inorganic material dimensions—following Adamic sin (and was not originally built-in by God). This is "the disordering of creation via natural evil" as posited by Dembski.[45]

Second, there is a very deep significance to the precise words used by Paul in verse 22 in reference to the created order. The key words are the simple phrase, "the whole creation." These words not only can be understood to support the previous assertion because of their general and all-inclusive character (they indeed do provide support), but they additionally carry embedded within them a meaning requiring a bit of textual exegesis, science, and philosophy to extricate.

---

43. Harrison, "Romans," 94. He says, "Christ spoke of the renewing of the world and called it a 'rebirth' (*palingenesia*, Matt 19:28)."
44. Blackman, "Romans," 784.
45. Dembski, "End of Christianity," 146.

The Meaning of "Whole Creation"

The central focus is upon Romans 8:22 and the deeper meaning of "the whole creation" (Greek *pas ktisis*). First, it shall be approached exegetically. The Greek word *pas* means "all or every."[46] The Greek word *ktisis* means "the sum total of what God has created."[47] Therefore, at the most basic level, *pas ktisis* is "every aspect [all] of the sum total [everything] of what God has created" (i.e., the entire created order—the natural universe; cf. Gen 1:1). In fact, the two words used together serve as a double reinforcement of the *totality* inference; Paul is emphasizing that everything outside of God himself (who is uncreated) and, by context, everything outside of the untainted heavenly realm and the unredeemable Satanic realm, is meant by "the whole creation."[48] Gerald R. Cragg concurs:

> We observe . . . Paul's sensitiveness to the pathos of nature's plight of subjection to futility; here he alludes more particularly to the *sorrow* of nature. He thinks of the sufferings of animals—the weak devoured by the strong—of the ruthless destruction of plant life, of natural catastrophes of all kinds; he listens, it is not too fanciful to suggest, to the cryings of the wind and the sea; and he receives an impression that all of nature is groaning in travail together, i.e., in all its parts. The whole created world is crying for release from pain . . . we [also] groan as we wait.[49]

Therefore, we conclude from an exegetical perspective that "the whole creation" refers to the entire natural universe.

Second, it shall be approached from the perspective of science, specifically, that of conventional physics.[50] Of what does conventional physics

---

46. Mounce, *Expository Dictionary*, "*pas*," 13 (G3650).

47. Mounce, *Expository Dictionary*, "*ktisis*," 146 (G2937).

48. See Scott, "Romans," 1154. He says, "The phrase ["the whole creation"] suggests to us pre-eminently though not exclusively the world of nature and of inanimate being; to Paul it probably stood mainly though not exclusively for the world of sentient being, including the spiritual forces in the unseen." We disagree with Scott on at least two counts. First, Paul clearly refers to both the inanimate and the sentient in the natural order; the force of the text is all-inclusive. Second, Paul could not have been referring to "the spiritual forces of the unseen" because the hosts of heaven are not groaning and the forces of darkness—though they may groan—are unredeemable (cf. Matt 25:41).

49. Cragg, "Romans" [Exegesis], 521.

50. We assume the veracity of conventional physics (as opposed to the variant forms of theoretical and speculative physics). Theoretical physics, by its very name and nature (often with the usage of such terms as "metaphysical virtual particle," "imaginary time,"

consider the entire natural universe, namely, the *whole creation*, to consist? Fischer, a research chemist by trade, says that the whole creation is "the entire realm of nature," which is "the realm of matter and energy, space and time."[51] Ross (astrophysicist) concurs with Fischer; he says that divine creation was "the beginning of space, time, matter, and energy."[52] He also asserts that the created order is "the universe of matter, energy, space, and time."[53] Max Tegmark, a physicist, even concludes that anything more or less than this four-dimensional reality would result in "dead worlds."[54] The key point is that scientific orthodoxy considers the natural order—the *whole creation*—by necessity to consist of the four basic dimensions of space, time, matter, and energy. It is often simply referred to as *space-time*.[55] It is therefore our assertion that the *whole creation* (Greek *pas ktisis*), of which Paul wrote about in Romans 8:22, is the essential equivalent to what modern science typically calls *space-time* (i.e., space, time, matter, energy).[56]

"dimensional bent state," and "quantum foam," etc.), often seeks to theoretically and creatively traverse beyond the self-defined natural parameters of science and into the realm of metaphysics, yet generally negating Deity and the supernatural. There is also another reason to move much more carefully in this area: namely, ordinary experience. See Saniga, "Geometry of Time and Dimensionality of Space," 131–43. In this essay, Saniga (who is himself a theoretical physicist) provides an astute precautionary comment: "Modern theoretical physics has entered the territory of scientific inquiry that lies so far from ordinary experience that there exists no rigorous observational/experimental guide to be followed. The only means physicists have at hand to navigate through this region is mere appeal of abstract and often counter-intuitive mathematical principles. Yet, sticking to mathematical beauty alone may not necessarily be a proper path leading to a discovery of new, more fundamental physical laws. For the history of science, and physics in particular, teaches us a very important lesson that novel, revolutionary ideas and paradigm shifts were almost always preceded and accompanied by new evidence from observations and experiments that had accumulated over particular periods" (131–32).

51. Fischer, *God Did It, But How?*, 54.
52. Ross, *Creation as Science*, 55.
53. Ibid., 69.
54. Tegmark, "On the Dimensionality of Space-Time," L69–L75.
55. Craig, *Time and Eternity*, 217–19. Craig explains the relationship of the dimensions of space-time: "[T]he universe did not spring into being at a point in a previously existing space. Rather space and time themselves came into being along with the universe [i.e., the "stuff"—matter and energy—of the universe], which implies creation out of absolutely nothing." Thus, space, time, matter, and energy are the dimensions of a natural order that once did not exist, but suddenly came to exist out of nothing previously existing (e.g., Romans 4:17).
56. See DiSalle, "Space-time," 867. DiSalle defines "space-time" as "a four-dimensional continuum combining the three dimensions of space with time in order to represent

The Matter of Corruption

Next, philosophy shall be applied. If time is a dimension of the created order (along with the three spatial dimensions) as is posited by conventional physics, just what is *time*? Plato called time "a moving image of eternity"; Aristotle said that it is "the number of movements in respect to the before and after; Augustine described time as "a present of things past, memory; a present of things present, sight; and a present of things future, expectation."[57] According to Craig:

> *Time* is that dimension of reality whose constituent elements are ordered by relations of *earlier than, simultaneous with,* and *later than* and are experienced by us as past, present, and future. This much, at least, is common property among almost all disputants in debates about the nature of time. Beyond that point, philosophers are deeply divided about the nature of time.[58]

The basic controversy in understanding time centers mostly around the debate as to whether time is *actually* tensed or tenseless (as opposed to simply *appearing* to be). In his 1908 essay, "The Unreality of Time," J. M. E. McTaggart labeled these concepts as A-Theory time (tensed time) and B-Theory time (tenseless time).[59] Interestingly, he also adds in a third concept called C-Theory time (ordered purpose time) that is not generally used in discussions about time because its nature is more kairological than chronological.[60] We will focus here upon A-Theory time and B-Theory time (however, *kairos* will come into play). Craig explains the difference:

> We are all familiar with tense as it plays a role in natural languages. But many philosophers hold that tense is an objective feature of reality as well, that things in time are really past, present, or future.

---

motion geometrically. Each point is the location of an event, all of which together represent 'the world' through time."

57. Earman and Gale, "Time," 920.

58. Craig, "Time, Eternity, and Eschatology," 596.

59. McTaggart, "The Unreality," 23–34. His precise terminology is A-Series and B-Series. McTaggart actually concluded that time, both in its tensed and tenseless understanding, is only a human perception and thus not real: "Nothing is really present, past, or future. Nothing is really earlier or later than anything else or temporally simultaneous with it. Nothing really changes. And nothing is really in time. Whenever we perceive anything in time—which is the only way in which, in our present experience, we do perceive things—we are experiencing it more or less as it really is not" (34).

60. For an in-depth discussion of C-Theory, see McTaggart, *The Nature of Existence,* 208–46. C-Series time is described as being something akin to B-Series, but shorn of its temporality, making it essentially a non-chronological concept.

> Other philosophers regard tense as purely mind-dependent: things in time are no more objectively "now" than things in space are objectively "here."[61]

In A-Theory time, events are not ontologically equal. The past no longer exists; the future has not yet come into existence; the present is all that is real. A-Theory is based on the concept of "temporal becoming," namely, things come and go out of being. Therefore, A-Theory time is often known as "presentism."[62]

In B-Theory time, events are considered to be ontologically equal. All events in time are equally real and the idea of temporal becoming is an illusion of the human mind. According to advocates of B-Theory, the tensed concepts of past, present, and future are simply relative properties (i.e., relative to when a person is living—e.g., what is future for a person in 1850 might be considered as past or present for a person living in 1950 or 2050, etc.). Contrary to the tensed *past*, *present*, and *future* labeling of A-Theory time, B-Theory time labels events as *earlier than*, *simultaneous with*, and *later than*. In this view, the tenseless relations are unchanging regardless of the relative position of an observer. Therefore, B-Theory time is often known as "eternalism."[63]

There are several reasons why we advocate the A-Theory of time over the B-Theory of time. Logically, as McTaggart himself realized, B-Theory is actually insufficient as a stand-alone entity. B-Theory time cannot exist without piggybacking somewhat on A-Theory time. B-Theory tenseless labels have no meaning apart from A-Theory tenses.[64] Moreover, and most importantly, the B-Theory becomes glaringly troublesome when applied to the Christian worldview.

First, there is the finite realism problem. Humanity, if indeed some form of comprehendible realism does exist (and we contend that it does), functions in a very real and objectively tensed environment. Unless everyone is grossly deceived by human consciousness (which would be an affront to the living God), human beings cannot traverse backward into yesterday or forward into tomorrow. People are bound by the present. This is part of what makes a finite creature (who is bound by chronological time)

---

61. Craig, "Time, Eternity, and Eschatology," 596.
62. Ibid., 596–97.
63. Ibid., 597.
64. McTaggart, "Unreality," 26–27.

different from the Infinite Creator (who is not bound by chronological time).[65] Furthermore, God seems to affirm a tensed form of time with the general *flow* of the Creation Week, and especially, by his direct words on Day Four (Gen 1:14): "Let there be lights in the firmament of the heavens to separate the day from night; and let them be *for signs and for seasons and for days and years*." This statement strongly supports the image of true and divinely-ordained temporal becoming.

Second, there is the eschatological problem. Holding an eternalist understanding of time drastically weakens the biblical tenet of the "not yet." It essentially relegates the divine promise of the Eschaton to an illusion. It means that the teaching of Christ for the church to watch and to wait is effectually irrational.[66]

Third, there is the problem of the permanent existence of evil. In accordance with the B-Theory of time, evil and death will never be fully eliminated from creation. As Craig states:

> [In B-Theory time] Evil's being destroyed amounts to no more than later portions of the space-time block's being free of evil. But the earlier parts infected with evil exist just as robustly as the later parts. The stain of evil on creation is indelible.[67]

Craig continues with the most unacceptable thought for the evangelical Christian:

> What this implies for Christ's crucifixion and resurrection is especially disturbing. In a sense Christ (or at least that temporal part of him) hangs permanently on the cross. The victory of the resurrection seems hollow, since the death and suffering are never really over and done away with. What kind of eschatology is it that cannot erase evil from the universe?[68]

Fourth, there is the problem of divine judgment. B-Theory posits that all material things—including humans—*perdure* (as opposed to *endure*). This means that all material things not only have spatial dimensions, but also internalized temporal dimensions. Logically, this means that

---

65. Note also that there is undeniable physical evidence of temporal becoming. All finite creatures are currently bound by the corollary effects of *aging*; yet, God remains immutable (e.g., Ps 102:26; Mal 3:6; Heb 13:8; Jas 1:17).

66. Craig, "Time, Eternity, and Eschatology," 609.

67. Ibid., 609–10.

68. Ibid., 610.

objects—including humans—never exist wholly at a specific time; they are sliced up temporally (i.e., *temporally extended*, stretched throughout *all* time, but never completely there at any time). Contrastingly, A-Theory holds that all material things—including humans—endure through time. This means that all material things have spatial dimensions, but not an internal temporal dimension as such. Accordingly, this means that, a rock or a human, does indeed exist completely in a specific point of time, but time is not constituent to its being in the same way as is space, matter, and energy (i.e., a rock or a human does not take up time or consist of time; rather, it exists in time); it is still a complete rock or person at all times and points of its existence no matter the age, year, or hour, yet continues to exist completely in time only at that present moment. It is in a true and constant state of *temporally becoming*. In B-Theory time, due to temporal extension, there are incredible and strange implications for the biblical doctrine of divine judgment. Craig explains the extreme absurdity:

> Perdurantism construes that the persons with whom we have intercourse to be mere parts of extended spatio-temporal "worms" which are not persons, for the whole extended object is neither self-conscious nor endowed with freedom, intentionality, and the like, properties essential to personhood. But then the eschatological doctrine of divine reward or punishment becomes incoherent. For the person who appears before God at the judgment is a different person from the person who did the good deed or committed the crime. But it would be immoral to accord praise or blame to a person who did not do the praise- or blameworthy acts. Alternatively, if one considers a temporal part large enough to have been located both at the time of the action and the time of judgment, then God must also accord praise or blame to an infinite number of other shorter parts who again are present at the judgment but not at the actions, since the extended part can be divided into all the smaller parts, each of which is a person. Thus, perdurantism makes a mockery of moral praise and blame.[69]

In essence, B-Theory means people are really not true and whole persons at any singular temporal point and God will have to be unjust at the final judgment. Moreover, an extension of this absurdity is as follows. Assume as per traditional Christian theology that God grants a person eternal life based on whether or not that person is in a state of saving grace at their physical death or at the Eschaton. With that in mind, suppose someone was

69. Ibid.

a faithful Christian for most of his life, but then becomes an apostate at the very end.[70] Under the A-Theory of time, the individual exists completely at that moment in time (as he has at every other moment) and he faces divine judgment in his actual condition. His relationship in Christ did not abide until the end and thus he faces judgment outside of grace. Meanwhile, B-Theory would promote a significant dilemma. Most of the temporal parts of the individual were in grace; only the final part was not. However, in the perdurantist view, no part is privileged over any other part. How can justice be done? Thus, the practical workings of B-Theory are not only absurd, but are grossly in conflict with an evangelical Christian understanding of the scriptural revelation, particularly as related to the atonement and eschatology.[71]

The Inundative Corruption Hypothesis

It is now appropriate to review the foundational assertions: [1] God created the entire natural universe in a "good" (perfect and pristine) state; [2] the phrase "whole creation" in Romans 8:22 refers to the entire natural universe that God created in the beginning (Gen 1:1); [3] conventional physics interprets the entire natural universe to consist of the four basic dimensions of space, time, matter, and energy; and [4] the dimension of time actually exists as an objective and tensed reality (past, present, and future). Having made these several assertions, how do they all tie together? What does it mean with regard to the relationship of Adamic sin to the obvious evidences of pre-fall death and corruption while keeping in mind the presumption of an antiquated Earth and universe?

We present the following hypothesis. It may be called *inundative corruption*. An explanation for the geological evidences of deep time degradation (including the presence of fossils indicative of millions of years of death) is posited in the drastic *auto-effect* that Adamic sin had upon the original perfect creation; namely, the instant corruption of all aspects of complete space-time, inclusive of not only the present and the future, but the past as well. According to this hypothesis, the pervasive effects of the Adamic fall (i.e. death, decay, corruption, etc.) completely and totally inundated everything that was made (space, time, matter, energy) and will

---

70. I am a Wesleyan Arminian. We believe that people can "fall from grace."
71. Craig, "Time, Eternity, and Eschatology," 611.

continue to have such effect until the final consummation of the Kingdom of God and the advent of the New Creation.

Perhaps the idea can be most illustratively presented by the image of a car windshield. The windshield represents the "whole creation"—space, time, matter, and energy. A rock strikes the center of the windshield with a mighty impact. The rock represents Adamic sin. Instantly, the *entire* windshield is completely inundated with a "spider-web" cracking effect. If we understand the point of impact to be the historical present of Adam and Eve (namely, the specific chronological time of the Adamic fall) with the past and future existing respectively on each side of the impact, then a pictorial image becomes visible. The full windshield (i.e., the "whole creation"—the whole space-time block), still being held together and remaining existent after the impact (i.e., the Adamic fall), would nonetheless be severely damaged throughout its entirety.

Another helpful illustration might be that of a hypothetical Edenic petroleum geologist taking core samples. If the geologist dropped a deep core drill five minutes before the fall, he would pull up pristine samples of all strata from the Precambrian basement to the Quaternary surficial deposition. However, being pristine, it would contain no fossiliferous material. If the same geologist then dropped another core drill five minutes after the fall in the very same proximate location, he would retrieve similar samples of all the same strata, but this time it would be filled with fossils and other signs of degradation. It would have been the same Earth and universe, but now entirely corrupted and thus drastically changed. (In fact, even that first core would then be bearing of fossiliferous content!)

God the Creator, who is not bound by chronological time (nor space, nor matter, nor energy), who is completely self-existent apart from the space-time block—but who interacts with his whole creation at his own will and for his own purpose (Greek *kairos*)—allowed the drastic effect of humanity's sin to become instantly and completely corrupting of the previously perfect existence. In so doing, humanity was enabled to glimpse and experience the consequential severity of life in rebellion against God. A perfect existence was absolutely real prior to the fall. Yet, the currently corrupt existence became absolutely real at the immediate instant of the fall.

In effect, Adamic sin drastically impacted all of *created reality*; that is, all of space, matter, energy, and time—including that of the chronological past (as well as the chronological future). Even the stratigraphic timeline of the Earth now reveals corruption and death whereas it did not before.

The Matter of Corruption

Like the change of a television channel, existence was immediately different. While a basic semblance of original creation remained, all things at once reverted from perfection to the degraded state of fallenness, as if what we observe now is all that creation ever was. Humanity, animals, trees, rocks, the Earth, the moon, the sun, all galaxies near and far—everything that exists became inundated and thus twisted, marred, and corrupted as consequence of the fall of Adam. This was an instantaneous, drastic, and authentic change of reality. As a result, what we presently observe in the empirical evidences of cosmic and geohistory is (now) *actually what really happened* from the beginning of creation. From our natural perspective, the creation has always been fallen. In fact, the only extant evidence of the existence of a perfect original creation (pre-reality change) is the scriptural record. Thankfully, that same record is also the evidence of a coming perfect New Creation, which will be a complete reversal of the effects of the fall (cf. Rev 20:11; 21:1–5).[72]

Unlike with the standard Irenaean theodicy of Old-Earth Creationism, this hypothesis adequately preserves the classic Augustinian understanding that the creation was originally perfect and that death (i.e., the ultimate manifestation of corruption) in all of its invasive forms is the final enemy (1 Cor 15:26) and will be ultimately abolished (Rom 8:19–21; Rev 20:14a; 21:4). According to Henry Morris III:

> Some have suggested that all living things were originally designed by God to die, that over the millions of years in which animal and prehuman life was developing, death played a perfectly natural role in the creation. Some have even taught that the death which God threatened Adam with was a "special" kind of death that applied to humans . . . If there were eons of pain, suffering, and death *before* the awful rebellion of Adam brought "death" into the world, then the suffering of our Lord Jesus becomes unnecessary. If the "wages of sin" is nothing more than some sort of spiritualized distance from the Creator, then the entire burden of sin becomes nothing more than a mental attitude. Heaven and hell are what you make of it. Twisting the words of Scripture so that Christ's physical death had no meaning is a terrible heresy.[73]

We agree completely with Morris. Any viewpoint that supports the existence of death and corruption in the created order prior to human sin by

---

72. Wesley, "New Creation," 500–510.
73. Morris, "Issues of Death," 22.

the original design of God is not only a terrible heresy (per Morris), but also blatantly naturalistic and non-Christian (per Schaeffer). However, to deny the strong postulations of modern science when not demanded by a responsible interpretation of Holy Scripture (as, we believe, is the common practice of Young-Earth Creationism concerning the matter of deep time) is neither wholesome nor resolving of the dilemma. Our hypothesis, on the other hand, does offer resolution. In addition to preserving the classic evangelical understanding of a "very good" original creation (contra Ross), it also preserves the common scientific interpretation of the geologic and fossil record that indicates the extreme antiquity of the Earth and shows evidence of death and corruption prior to the advent of humanity (contra YEC).

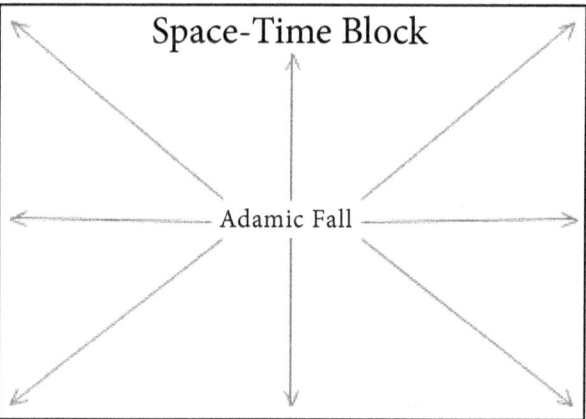

Figure 1: The Inundative Corruption of *pas ktisis*

3

The Foundations of Correlation

THE ASSERTIONS REQUIRING CORRELATION

IT IS THE ASSERTION of Old-Earth Creationism that God created the Earth and then made it into an inhabitable environment over the course of a "Week" of *epoch* long Creation "Days." It is the assertion of modern science that the Earth is 4.6 billion years old and has reached such an age by passing through a number of geologic periods that are differentiated by stratigraphic, paleontological, and other empirical markers. Therefore, it seems very logical that if one holds to the veracity of these two basic assertions, then the long Creation Days of Genesis and the geologic ages of modern science can and should be effectively correlated with one another in some cohesive and systematic manner. Unfortunately, and quite astonishingly, this is a project that has not yet been adequately accomplished. It is often claimed that such correlation simply cannot be done because of the supposed irreconcilable discrepancies existing between the order of God's creative work in Genesis and the postulations of modern science. For instance, Stephen J. Godfrey, a Christian and a paleontologist, says, "Many believing Christians have experienced crises of faith and personal rejections when they have chosen to accept an account of origins that is based on reasoned interpretation of centuries of scientific observation, because this account does not coincide with a literal interpretation of Genesis."[1]

---

1. Godfrey and Smith, *Paradigms on Pilgrimage*, 11. Godfrey and Smith convey

Perhaps the lack of effort to find a consistent reconciliation is due to this very kind of thinking: namely, the "literal" record of Genesis 1–2 *cannot* be found commensurate with "an account of origins that is based on reasoned interpretation of centuries of scientific observation" because the two do not *coincide*. Conrad Hyers makes this claim:

> Biblical affirmations are in harmony with the science of every period and culture, not because they have been harmonized by enterprising souls, but precisely because they have little to do with science. All attempts at synchronizing the Genesis materials with materials from the various sciences presuppose that they are in some way comparable, but they are not. Trying to compare them is not even like comparing apples and oranges. It is more like trying to compare oranges and orangutans.[2]

It is thus the view of both Godfrey and Hyers that Genesis and nature not only do not coincide, but they are not even comparable. Yet, despite such skepticism toward a concordantist approach and specifically, a correlation model, it seems that the search for truth requires the effort.[3]

THE SIGNIFICANCE OF THE MATTER

While this may seem at first glance to be quite an insignificant matter, it actually has very significant ramifications. First, such a correlation would provide substantiation that the Genesis 1–2 text records a series of actual historical events that are set in space-time. This lends credence to the notion that the biblical creation events themselves are not merely some mythological formulation with a small kernel of truth hidden somewhere deep inside, but rather events that are empirically verifiable and objectively

---

their non-concordantist view with this statement: "As Christians, we are not obligated to try to put new wine into old wineskins. There is nothing to be gained by trying to force all of the information that is now available to us about the nature of the universe into a world view that must by now be recognized as phenomenological, not objective" (192). Godfrey is certainly correct that the new should not be *forced* into the old, for truth by its very being does not have to be forced. However, if something is true, it means that it corresponds to reality and remains coherent to itself in every way. Therefore, the Genesis text, if it is indeed the revelation of God, must be concordant to all aspects of the creation of God, if both Scripture and nature are true.

2. Hyers, *Meaning of Creation*, 31.

3. Ramm, *Christian View*, 25. He says that "the two books of God must eventually recite the same story."

real (any use of phenomenological language to communicate the events notwithstanding).

Second, such a correlation would provide another evidential reason for practitioners of modern science to give serious consideration to the truth claims of the Gospel of Jesus Christ. Davis A. Young and Ralph F. Stearley speak straightforward to this important matter of Christian evangelization:

> [T]he worldwide professional geological community, consisting of tens of thousands of geologists, both Christian and non-Christian, is totally convinced of the vast antiquity of our planet. They are also convinced that sedimentary rocks were formed by a vast variety of processes, including deposition by wind, rivers, ocean waves and glaciers along coasts, in the deep ocean, in river valleys, on desert floors, rather than by a single, global, catastrophic Flood. Mainstream geologists are persuaded that young-Earth creationism and Flood geology, despite the façade of scientific sophistication with which they are presented to a geologically naïve public, are deeply flawed both scientifically and biblically. Christian geologists are especially concerned that persistent efforts by well-intentioned believers to gain acceptance for young-Earth creationism as a viable alternative to mainstream geological thinking will result in deepening alienation of the scientific community from evangelical Christianity, a trend that does not bode well for evangelization of scientists.[4]

They add that it is important "to demonstrate to non-Christians who may understandably entertain the false impression that Christianity entails commitments to a young-Earth and a global deluge that such commitments are by no means inherent to Christian faith."[5] It is the biblical mandate of every Christian to take seriously the evangelistic exhortation of Jesus (Matt 28:19–20—"Go therefore and make disciples of all nations, baptizing them in the name of the Father and of the Son and of the Holy Spirit, teaching them to observe all that I have commanded you.") and the apologetic exhortation of Peter (1 Pet 3:15—"Always be prepared to make a defense to anyone who calls you to account for the hope that is in you, yet do it with gentleness and reverence."). A definitive correlation between scripture and

---

4. Young and Stearley, *Bible, Rocks and Time*, 23.

5. Ibid. Note: We dispute the notion that the concept of a young-Earth and the concept of the global Deluge must be accepted or rejected in tandem. We believe that the two entities should be viewed separately.

nature would provide a very powerful compulsion toward effective Christian evangelism and apologetics.

Third, such a Creation Days-geologic column correlation would have great additional faith building and encouragement value for those who are already Christian believers. Young and Stearley claim that it is also very important "to convince Bible-believing Christians that the Earth really is extremely old and to show them that acceptance of such a belief need not be in any way a threat to their Christian faith."[6] We believe, in fact, that it would give additional support to the confidence of believers. Not everything posited by modern science is commensurate with the Christian faith and an evangelical interpretation of Holy Scripture, but much of it can be understood to be properly compatible.

THE TWO PRESUPPOSITIONS

We believe that it is necessary to provide an example schematic that shows a definitive correlation between the Creation Days of Genesis and the geologic timescale of modern science. Our correlation model is predicated upon two basic presuppositions (which we believe to be very evangelical). First, God did his creative work with the supreme end being that of making an inhabitable living environment for the crown of his creation, humanity. From the very beginning, everything has been moving toward this anthropic end.

Second, God's revelation in Genesis 1–2, while an accurate historical representation, is not absolutely exhaustive in terms of what he created on each Creation Day. (The Genesis text records the high points of God's creative work, within a framework of Mosaic phenomenology, leading to the divine completion of an inhabitable living environment for people.) With the acceptance of these presuppositions, Scripture and nature fall much more cohesively and coherently into place with one another.

THE JUSTIFICATION OF THE ASSERTIONS

Before describing the correlation model, it is appropriate to take a quick step back and discuss the two key assertions that we seek to correlate, namely the assertion of Old-Earth Creationism and the assertion of modern science.

---

6. Ibid.

THE FOUNDATIONS OF CORRELATION

Old-Earth Creationism asserts that the Creation Days of Genesis are epoch days and that God orderly and progressively made all that was made over long periods of time. Meanwhile, modern science asserts that the Earth is extremely old and that it has passed through a number of evident and very lengthy geologic ages.

The Assertion of Old-Earth Creationism

This inevitably brings to the forefront the matter of interpreting the word "day" (Hebrew *yom*) in the Genesis text. From a scriptural perspective, the assertion of an Old-Earth hinges on how *yom* is understood. According to both Francis Brown and William Mounce, the word *yom* can have any of several meanings depending on its usage and intention. First, it can mean the day as opposed to night (i.e., day time). Second, it can mean a basic division of time (e.g., a day's journey, a specific holy day, etc.). Third, it can mean the time of God's judgment (i.e., the day of Yahweh). Fourth, it can mean a plurality of time (e.g., all of a person's days, advanced in days, etc.). Fifth, it can mean an indefinite period of time that can be long or ancient (e.g., days of old).[7] While Young-Earth Creationism typically interprets the *yom* of Genesis 1 to refer to denotation two (a basic division of time often considered to be the rough equivalent of a solar day), Old-Earth Creationism typically interprets the same *yom* to refer to denotation five (often considered to be an indefinite, yet finite, epoch of time). Since either denotation can be semantically correct, the key to discerning which meaning to claim for Genesis 1 must be determined from an integration of textual-literary context, theological context, and/or from any extra-biblical information that might reasonably and responsibly apply.[8]

Young-Earth Creationism holds to the solar day view based primarily on textual design (i.e., semantic, syntactical, etc.) considerations. Basically, since *yom* is used in combination with "evening" (Hebrew *ereb*) and

---

7. Brown et al., *Hebrew and English Lexicon,* "day," 398–99 (H3117) and Mounce, *Expository Dictionary,* "day," 157–58. For further, see Wilson, *Wilson's Old Testament Word Studies,* 109; and Vine et al., *Vine's Complete Expository Dictionary,* 54.

8. Moreland, "Introduction," 11. Moreland states: "Christians have a special intellectual and moral obligation to follow Augustine's advice: we have a duty, he said, to show that our Scriptures do not contradict what we have reason to believe from reliable sources outside them. In short, Christians have the obligation and privilege of developing and propagating an integrated Christian worldview."

"morning" (Hebrew *boqer*) on the first six days,[9] and since each *yom* is used with specific numerics (the first day uses the cardinal form "one"; the other days use ordinal forms "second," "third," etc.), this construction is understood to imply the intended meaning to be solar days.[10] Jacques B. Doukhan explains the classic Young-Earth idea:

> Regarding the nature of the days of the creation week, the text is quite clear and explicit. The text does not imply that they are symbolic or cosmic, but it gives us enough clues about the author's intention to refer to days that are of the same temporal nature as our human days ... Although humans are present on the sixth day, which is also qualified with the expression "evening and morning," the fact that they have been created within the day implies, indeed, that only the seventh day was their first and only full day of the creation week. Only the seventh day was the day they experienced totally, from sunset to sunset. For this day, we do not need, therefore, the specification "evening and morning." For the other six days, on the other hand, humans are totally or at least partially absent, and therefore the author feels necessary to specify "evening and morning" to make it clear and emphasize that these days are of the same nature as our human days.[11]

Young-Earth Creationism asserts that there is overwhelming textual substantiation for literal, 24-hour creation days. John MacArthur strongly concurs: "I am convinced the correct interpretation of Genesis 1–3 is the one that comes naturally from a straightforward reading of the text. It teaches us that the universe is relatively young, albeit with an appearance of age and maturity, and that all of creation was accomplished in the span of six literal days."[12]

On the other hand, Old-Earth Creationism understands that the "creation week" consists of six (or seven) periods of indefinite length. Thus, each *yom* is understood, from the human perspective, to be a long period of time that is described figuratively in the text, from the God-perspective, as a "day." J. Oliver Buswell, Jr. explains this classic Old-Earth idea:

> I believe that Moses uses the word day in the first chapter of Genesis in a figurative manner ... When you think of your Bible, the

---

9. Sarfati, *Refuting Compromise*, 81.
10. Ibid., 76.
11. Doukhan, "Genesis Creation Story," 25–26.
12. MacArthur, *Battle for the Beginning*, 29–30.

phrases "the day of the Lord," "that day," "the last days," [and] "the last hour" come to mind. Christ said, "It is the last hour," when He was speaking to His people there in the days of His flesh. "It is the last hour"—and that hour has already been over 1900 years long. The figurative use of the words for parts of time, the word for "days" ought to be familiar from the Scripture. We should not have any great difficulty with this concept.[13]

Ross, following this same line of thought, asserts, "Old-universe Christians say the text allows ample room, with no compromise of biblical inerrancy, for creation days of longer duration and even for a cosmic origin date of just over 10 billion years."[14]

The Assertion of Modern Science

Modern science makes several key claims relating to antiquity. First, science currently holds that the universe is 13.72 ± .12 billion years old.[15] This figure is one result of an extensive five-year satellite study of the cosmos and conducted cooperatively by NASA and Princeton University from August 10, 2001 to August 9, 2006.[16] Gary F. Hinshaw et al., explains:

> The Wilkinson Microwave Anisotropy Probe (WMAP) is a Medium-Class Explorer (MIDEX) satellite aimed at elucidating cosmology through full-sky observations of the cosmic microwave background (CMB). The WMAP full-sky maps of the temperature and polarization anisotropy in five frequency bands provide our most accurate view to date of conditions in the early universe. The multi-frequency data facilitate the separation of the CMB signal from foreground emission arising both from our Galaxy and from extragalactic sources. The CMB angular power spectrum derived from these maps exhibits a highly coherent acoustic peak structure which makes it possible to extract a wealth of information about the composition and history of the universe, as well as the processes that seeded the fluctuations.[17]

---

13. Buswell, Jr., "Creation Days," 10.

14. Ross, *A Matter Of Days*, 18. Note: This is the general relative age of *our solar system*.

15. Hinshaw et al., "Five-Year Wilkinson Microwave Anisotrophy Probe (WMAP) Observations," 242. See also, Lambert, *Field Guide to Geology*, 16.

16. Ibid., 225–28.

17. Ibid., 225.

Second, science posits that the Earth is 4.567 billion years old.[18] Precambrian rock (namely, all rock prior to the Phanerozoic strata, of which the Cambrian Period is the first geological system) is quite sparse prior to the 3.8 billion year point (i.e., the relative ending boundary of the Hadean Eon, which is understood to have been the period of molten and "hellish" conditions of the primordial Earth). Therefore, at this time, the most effective way of providing an approximate age of the Earth is two-fold: [1] by radiometric dating of interbedded volcanic rocks and plutonic rocks, particularly of the mineral zircon, which originate from the deepest recesses in the Earth,[19] and [2] by considering terrestrial meteorites and retrieved moon rocks to be part of the initial Solar System accretion and equating their age with that of the Earth.[20] According to Ogg, "It is now well documented that the condensation of solid material to form the terrestrial planets in our Solar System occurred at 4.567 Ga (or T_0, and that accretionary processes continued for ~30–100 myr thereafter, including the formation of the Moon as a result of the glancing impact of a Mars-size planet called *Theia*, about 40 myr after T_0."[21]

Third, science divides the time of Earth's existence into time sequential geologic ages. Ogg states, "The time scales are based on regional and global geologic mapping, which establishes relative ages of surfaces delineated by superposition, transaction, morphology, and other relations and features."[22] He explains the logic undergirding the sequential system:

> Geologic time and the observed rock record are separate but related concepts. A geologic time unit (geochronologic unit) is an abstract concept measured from the rock record by radioactive

---

18. Ogg et al., *Concise Geologic Time Scale*, 24.

19. For an excellent and very balanced treatment of the subject, see Wiens, "Radiometric Dating." In the article, he also presents a comprehensive list of resources for those interested in further research. For a contrasting opinion, see Froade, Jr., "Radiometric Cherry-Picking." He argues against radiometric dating by presenting several inconsistency anomalies. However, he fails to present the overwhelming level of its consistency.

20. Ogg et al., *Concise Geologic Time Scale*, 23–24. This will be the methodology used concerning Precambrian rock "until packages of strata and associated global events could be recognized and correlated by the intrinsic features of their geologic history rather than simply by numerical dates" (24). See also Murck, *Geology*, 62; and Lambert, *Field Guide to Geology*, 16–18. Murck provides an excellent essay on how the age of the Earth is determined.

21. Ibid., 26–27. See Lambert, *Field Guide to Geology*, (18–19) for a clear and concise explanation of cosmic accretion theory.

22. Ibid., 13.

decay, Milankovitch cycles or other means. A "rock-time" or chronostratigraphic unit consists of the total rocks formed globally during a specified interval of geologic time. Therefore, a parallel nomenclature system has been codified—geologic-time units of period/epoch/age that span the rock-record units of system/series/stage. The period/systems are grouped into eras/erathems within eons/eonthems, respectively.[23]

A basic understanding of stratigraphy is important to understanding rocks and geologic time.

Stratigraphy is the scientific study of rock strata.[24] Barbara Murck provides three foundational principles of stratigraphy that led to the development of the standard geologic time scale. These principles are as follows: [1] the principle of original horizontality (waterborne sediments deposit as horizontal layers), [2] the principle of stratigraphic superposition (a sedimentary rock layer is younger than the layer below it and older than the layer above it), and [3] the principle of crosscutting relationships (rock is always older than any other cutting or disruptive feature that disturbs it).[25] Alan M. Cvancara provides two additional principles used by field geologists: [4] the principle of inclusion (fragments of older rocks can be enclosed within younger rocks) and [5] the principle of faunal progression (fossil assemblages occur in a definite order in the vertical rock sequences).[26] These time-honored and accepted principles of stratigraphy provide geologists with the capability to make determinations as to the relative age (i.e., the age of rock relative to other rocks) of rocks and rock formations.

What does stratigraphy reveal about the Earth? First, it is clearly observed that each layer of strata contains "specific assemblages of fossils."[27] This observation is very important to the chronological ordering of rock sequences. Second, it is clearly observed that there is a definitive and orderly stratigraphic correlation.[28] This observation is important in the relative

---

23. Ibid., 4.

24. Murck, *Geology*, 49.

25. Ibid., 49–51.

26. Cvancarra, *Field Manual*, 122–123. See also Eerdman, "Stratigraphy and Paleontology," 3–6, 11. Eerdman strongly justifies using fossil assemblages in determining the relative age of rock.

27. Murck, *Geology*, 52. Also, for a complete account of this "specific assemblages" phenomenon as it exists in the Grand Canyon, see Eerdman, "Fossil Sequence in Clearly Superimposed Rock Strata," 13–17.

28. Ibid., 53.

dating of rock sequences found in different locations on Earth. As Murck explains:

> One of the greatest successes of nineteenth-century science was the demonstration, through stratigraphic correlation, that sequences of rock strata are the same on all continents. This meant that a gap in the stratigraphic record in one place could be filled by evidence from somewhere else. Through worldwide correlation, nineteenth century geologists assembled the geologic column, the succession of all known strata fitted together in chronologic order on the basis of their fossils or other evidence of relative age.[29]

Thus, the rock and fossil assemblages relate directly to age and time. As C. Scott Southworth of the U. S. Geologic Survey in Reston, VA once remarked, "The entire history of the world is preserved in the rocks."[30] The phrase "the entire *history* of the world" implies the passage of time; the phrase "*preserved* in the rocks" implies the existence of a record. There is a definitive relationship between the geologic column (i.e., the extant natural rock record) and geologic time (i.e., the history of the world). Murck explains:

> Standard names are now used worldwide for the subdivisions of the geologic column. The geologic column is divided into four major time divisions called eons. The eons, which all have Greek names, are the Hadean ("beneath the Earth"), Archean ("ancient"), Proterozoic ("early life"), and Phanerozoic ("visible life"). The eons encompass hundreds of millions to billions of years. They are divided into shorter spans of time called eras. Eras are most useful into dividing up the Phanerozoic eon because they are defined by fossil assemblages; fossils are absent or very rare in rocks of the earlier eons. The three eras of the Phanerozoic eon are the Paleozoic era ("ancient life"), Mesozoic era ("middle life"), and Cenozoic era ("recent life").[31]

Rusbult adds this supportive thought: "Evidence from a wide range of fields—including the study of sedimentary rocks, the geological column, the fossil record in geological context, coral reefs, and seafloor spreading (caused by continental drift) with magnetic reversals, plus (in non-geological fields of

---

29. Ibid.

30. Southworth. Unable to locate the original source of this statement, its attribution was confirmed by a personal email from him on February 17, 2011.

31. Murck, *Geology*, 53.

The Foundations of Correlation

science) radioactive dating, genetic molecular clocks, the development of stars, starlight from faraway galaxies, and more—indicates that the earth and universe are billions of years old."[32] Cvancara concurs:

> In the seventeenth century, Biblical scholars stipulated that prehistoric time extended back to about 6,000 years; this figure prevailed until the nineteenth century when several dating attempts, based on scientific principles, pushed the beginning of prehistoric time back millions of years. Geologists have since demonstrated that billions, not thousands or millions, of years encompass prehistory. Prehistoric time since Earth's origin is often referred to as geologic time.[33]

The bottom line is this: it is the comprehensive understanding of modern science that the Earth and universe are not only significantly antiquated, but that the natural history of the Earth includes numerous very long time periods that are successively indicated by certain specificities of the geologic record and supported by index fossil evidence demonstrative of the great passage of geologic time. Moreover, modern science further holds that the geologic column is the record of geologic time, and that geologic time is *deep time*.[34] John McPhee, who is credited with coining the term "deep time," puts it in clear perspective:

> With your arms spread wide again to represent all time on earth, look at one hand with its line of life. The Cambrian begins at the wrist, and the Permian Extinction is at the outer end of the palm. All of the Cenozoic is in a fingerprint, and in a single stroke of a medium-grained nail file you could eradicate human history. Geologists live with the geologic scale.[35]

Geology, in fact, is predicated upon the geologic timescale (i.e., the geologic column).

---

32. Rusbult, "Do we have evidence?"
33. Cvancarra, *Field Manual*, 116–17.
34. See McPhee, *Basin and Range*, 133. He says: "People think in five generations—two ahead, two behind—with heavy concentration on the one in the middle. Possibly that is tragic, and possibly there is no choice. The human mind may not have evolved enough to be able to comprehend deep time. It may be only able to measure it."
35. Ibid., 132.

THE REALITY OF THE GEOLOGIC COLUMN

Despite the strong evidential claims of modern science, there are some who deny the existence of the geologic column.[36] There are also many others who deny the common interpretation of the geologic column.[37] It is necessary to speak to those matters.

The Existence of the Geologic Column

First, it is important to substantiate that the geologic column truly does exist. This is really only an issue because it has been claimed that the column is only a reality in science textbooks and not in the actual natural order. Henry M. Morris and Gary E. Parker make this comment:

> The column is supposed to represent a vertical cross-section through the earth's crust, with the most recently deposited (therefore youngest) rocks at the surface and the oldest, earliest rocks deposited on the crystalline "basement" rocks at the bottom. If one wishes to check out this standard column (or standard geologic age system), where can he go to see it for himself? There is only one place in all the world to see the standard geologic column. That's in the textbook![38]

Meanwhile, John Woodmorappe has come to a similar conclusion: "It [the geologic column] remains primarily an invention of the uniformitarian imagination, and a textbook orthodoxy."[39] He also adds:

> There is no escaping the fact that the Phanerozoic geologic column remains essentially non-existent. It should be obvious, to all but the most biased observers, that it is the anti-creationists who misrepresent the geologic facts. The geologic column does not exist to any substantive extent, and scientific creationists are correct to point this out.[40]

---

36. Woodmorappe, "Geologic Column," 77–82.

37. See Sewell, "Uniformitarianism and the Geologic Column," 1–3; Froede, Jr., *Geology by Design*, 8–12; Tyler and Coffin, "Accept the Column, Reject the Chronology," 53–68. It should be noted that Froede tries to differentiate between the reality of the *actual rock record* and the reality of the geologic column (9).

38. Morris and Parker, *What Is Creation Science?*, 230.

39. Woodmorappe, "Geological Column," 3.

40. Ibid., 8.

The Foundations of Correlation

We respond with some Young-Earth advocates who, though they disagree with the common explanation of the column, do indeed affirm its legitimate physical existence. For instance, David J. Tyler and Harold G. Coffin state:

> This story has been repeated many times. Local sections consistently confirm the geological column concept. Near the base of the fossiliferous rock sequences (Paleozoic rocks), the dominant fauna are marine, with trilobites, nautiloids, graptolites and certain forms of coral. Toward the upper Paleozoic are major coal-bearing strata. Above these are red-beds largely devoid of macrofossils, followed by a variety of terrestrial and marine strata, often bearing witness to the presence of dinosaurs. This culminates in chalk, a very extensive and distinctive rock type, with its own suite of characteristic fossils. Above this are Tertiary beds, typically less persistent, with fossil mammals and birds (but no dinosaurs). This general sequence comes close to being a global phenomenon.[41]

Harold W. Clark, a geologist—also a Young-Earth creationist—concurs. He makes the following statement about the geologic column:

> It is important to observe that these were real, not imaginary, rock divisions and that they actually occurred in this order... The rocks above the Paleozoic were called Mesozoic, or middle life. This group contained fossils of higher groups of animals than did the Paleozoic. Then above the Mesozoic was a different type of rocks, with their characteristic fossils. This group was named Tertiary. Above these the Quaternary completes the geological column.[42]

Clark provides a few more candid thoughts. He says, "No one who has observed them can deny that there is order and system to the stratified rocks."[43] He also asserts, "The geological record is valid."[44] It should be noted that Clark spent a great deal of time as a *field* researcher.

Meanwhile, here is the thinking from the Old-Earth camp. Daniel E. Wonderly explains the reason why the geologic column is described as such: "We speak of such a sequence of layers of rocks as a 'column' because we see it as such when looking at the drilling cores which are removed

---

41. Tyler and Coffin, "Accept the Column," 58–59.

42. Clark, *Genesis and Science*, 61. Note: The Tertiary and the Quaternary are periods of the Cenozoic Era.

43. Ibid., 76.

44. Ibid., 80.

when drilling through the layers, and also because it is convenient to picture it as a narrow column in books and journals, thus saving space."[45] Wonderly's description is particularly revealing because it emphasizes that the common term used (column) ties the phenomenon to direct observation (drilling cores). In other words, the geologic column is something that is plain and obviously seen and quite tangible. It cannot be reasonably denied without rejecting empirical realism.

Moreover, just how expansive is the column? Are Tyler and Coffin correct in claiming that it is "close to being a global phenomenon"? Derek V. Ager waxes prosaically on this issue:

> In 1957 I had the good fortune to visit the geologically exciting country of Turkey, to look at some of the local Mesozoic rocks and their faunas. I was taken by a Turkish friend to visit a cliff section in Upper Cretaceous sediments near Sile on the Black Sea coast ... What I was looking at was identical with the 'White Cliffs of Dover' in England and the rolling plateaus of Picardy in France, the quarries of southern Sweden and the cliffs of eastern Denmark ... We have long known, of course, that the White Chalk facies of late Cretaceous times extended all the way from Antrim in Northern Ireland, via England and northern France, through the Low Countries, northern Germany and southern Scandinavia to Poland, Bulgaria and eventually to Georgia in the south of the Soviet Union. We also knew of the same facies in Egypt and Israel. My record was merely an extension of that vast range to the south side of the Black Sea.[46]

Though Ager is here only referring narrowly to Mesozoic strata (even more specifically to Upper Cretaceous sedimentation), his point should be well taken. He provides this summative thought, "At certain times in earth history, particular types of sedimentary environment were prevalent over vast areas of the earth's surface. This may be called the 'Phenomenon of the Persistence of Facies.'"[47] His point is that sedimentary deposition—translating over time into stratification (i.e., geologic column)—is remarkably consistent all over the planet. This affirmation of the column should not be misunderstood to mean that all Phanerozoic strata exist everywhere a core is removed. There are many, many places where *unconformities* (namely, missing sections or irregularities in the rock record) exist. This is, however,

45. Wonderly, *God's Time-Records*, 49.
46. Ager, *Stratigraphical Record*, 1.
47. Ibid., 23.

The Foundations of Correlation

an expectation of basic geology. Unconformities and other lithographic deformations are easily explained by such phenomena as the cessation and recurrence of sources of sedimentary deposition (e.g., streams, rivers, seas, wind, etc.), normal erosion, faulting, and folding actions, etc.,[48] as well as occasional catastrophic events (e.g., turbidity currents, earthquakes, etc.).[49] Over the course of deep time, there are many natural forces that can effect what *visibly* endures within the geologic chronicle. As Murck points out so well, "[U]nconformities are important because they can represent very long periods of time that are missing from the sedimentary rock record."[50] In contrast, Woodmorappe claims that less than 1 percent of the Earth's surface embodies the complete geologic column.[51] Further, he claims that unconformities in the record "do not engender confidence in the reality of the column."[52] Woodmorappe is mistaken. He seems to misunderstand that gaps or discontinuities in the column do not negate the elements of prehistoric time, only the physical record of that time at that particular location. Yet, even the gaps in the record at one location can be filled in from a more complete record at another location.[53] Richard Ritland says:

> Upon reflection it is obvious that substantial deposition cannot normally occur over the entire earth's surface at the same time. There must always be a source for every particle of gravel, sand, lime, silt or lava which is deposited. It follows that no single interval or moment in earth history can be everywhere represented by deposits. There are no universal formations . . . Moreover, since there is much evidence to support the view that no single region has continuously received deposits from the time of the creation of the planet until now, in developing a complete "geologic column" for the world it is necessary to attempt to correlate strata so that any level where non-deposition or an erosional break has occurred may be represented by rocks from other areas laid down at the time of the missing interval. This introduces a subjective element, but fortunately there are hundreds of regions with extensive series of superimposed strata representing major segments

---

48. Murck, *Geology*, 50–51. See also, Eerdman, "Fossil Sequence," 1–2.
49. Ager, *Stratigraphical Record*, 55–70.
50. Ibid., 51.
51. Woodmorappe, "Geologic Column," 4–5.
52. Ibid., 7.
53. Murck, *Geology*, 53.

of the column, and there are diverse criteria for correlation available so that correlations necessary may generally be quite firmly established.[54]

The geologic column is simply not expected by modern science to be locally complete in all places. According to Glenn R. Morton:

> There is no place on earth that has sediments from every single day since the origin of the earth. No geologist would require this level of detail from the geologic column. But if there are sediments left at a given site once every hundred thousand years or so, then at the scale of the geological column, the entire column would exist. There would still be erosional surfaces contained in that column and that would mean that some days left no sediment at a given location to mark their existence.[55]

It is Morton's suggestion that the best way to clear up the matter is to examine the resources used by the major oil companies for exploration.[56] The most widely used technical resource is the *Stratigraphic Database of Major Sedimentary Basins of the World*.[57] This is the primary and most complete database concerning the geologic column and is used as a standard research tool by the oil industry throughout the world. Using the *Stratigraphic Database* as his source, Morton selects twenty-six very specific and somewhat evenly distributed intercontinental locations where petroleum geologists have documented in great detail the complete geologic column; each of the samples is inclusive of significant strata from the Cambrian Period to the Quaternary Period (the entire Phanerozoic sequence). Furthermore, as an example, Morton also specifically breaks down and describes in very precise detail the stratification specifically located in the Williston Basin of North Dakota (which is one of the twenty-six aforementioned locations).[58] He provides a very strong technical and observational basis for affirming the existence of the geologic column as posited by conventional geology. In other words, the geologic column is physically real.

---

54. Ritland, "Historical Development: Part I," 61.
55. Morton, "Geologic Column," 2.
56. Ibid., 12. Their primary motive is monetary profit and not a particular ideology.
57. Robertson Group, *Stratigraphic Database*, 1989.
58. Morton, "Geologic Column," 3–11.

The Interpretation of the Geologic Column

This is a very significant matter. As Ritland says, "The crucial questions on the relationship of Genesis and geology, of religion and geological science nearly all hinge in some way on one's understanding of the meaning and significance of the geologic column."[59] It is important now to discuss how the geologic column should be interpreted and understood.

Typically, there are two primary schools of thought on this. Again, from the Christian perspective, they are divided along the age of the Earth issue. There is the *uniformitarian* view and the *diluvial* (a form of catastrophism) view.

According to classic uniformitarianism, the column represents vast ages of time. Each stratum is the physical result of many millions of years of sedimentary deposition and lithographic compaction. Therefore, each stratum "tells" the story of a long time period of Earth history. As the ages pass, other strata are laid down on top of the previous strata (and so on and so forth). This process of deposition continues on up to the present day. In a nutshell, uniformitarianism holds that the Earth is very old and that depositional processes have been going on more or less in a similar (uniform) fashion since the beginning of geologic time.[60] It should be noted that uniformitarianism does allow for the occurrence of local and regional (and sometimes, wider ranging) catastrophic actions, even episodic events.[61] Such occurrences (earthquakes, floods, meteors, etc.) are clearly evident at many places in the rock record.[62]

Contrastingly, according to standard diluvialism, the column largely represents the drastic effects of the Noahic Flood upon the Earth's crust. Advocates of such diluvialism hold to an overt catastrophist understanding. This means that instead of viewing the current terrestrial conditions as the result of long periods of slow processes, the conditions are understood as being the result of a sudden and drastic action caused by a worldwide catastrophe: the flood of Noah. Typically, standard diluvialism also holds to a young Earth. The basic idea is that since the Earth is less than 10,000 years old, the uniformitarian process (which requires millions and millions of years) could not have occurred. The narrow time constraints of Scripture

---

59. Ritland, "Historical Development: Part I," 59.
60. See Ritland, "Historical Development: Part II," 28–49.
61. See Ager, *Stratigraphic Record*, 73–84, esp. 83–84; also, *New Catastrophism*.
62. Wonderly, *God's Time Records*, 59.

will not allow for vast ages. Therefore, the best explanation for the crustal conditions of the Earth is diluvial catastrophism.[63] Sewell clarifies:

> [Young-Earth] Creationists suggest a totally different way of looking at the fossil record. We believe that almost all the sedimentary fossil-bearing rocks covering the earth's surface were deposited as a direct result of the Great Flood of Noah. Those small marine creatures at the lowest depths of the ocean when the Flood began were in a position to be quickly covered with the initial surge of the sediments. (These are the ones that evolutionists say must have lived first.) Other kinds of creatures such as birds, land-dwelling mammals, etc. could have escaped the earliest devastations of flooding. This is an easy, natural explanation for the separation of fossils into groups.[64]

He also makes this comment: "A Christian should realize that acceptance of this evolutionary dating scheme involves rejection of the Bible's clear teaching that the earth was created by God in six days, no more than 10,000 years ago. The naturalistic teachings of evolution and an ancient earth are based on the belief that parts of the Bible aren't really true."[65] It should also be noted that Young-Earth Creationism has devised its own version of the geologic scale based on Noahic Flood inundation and recession stages.[66]

There are clearly two differing viewpoints on how the geologic column should be interpreted. The issue may seem to come down to a simple matter of trusting Scripture or trusting science. Actually, that is not the case at all. As Ritland reminds the scholarly world:

> It has been demonstrated that the basic framework of the geologic column was founded by men with respect for Scripture, who, although not holding to conservative interpretations, opposed organic evolution. Anyone who reads the original literature will soon recognize that there was no conscious conspiracy on the part of these scientists to undermine the moral and religious authority of Scripture as sometimes has been charged. Completely apart from any merits or weaknesses, the geologic column is the result of an attempt by conscientious scientists to construct to the best

---

63. Sewell, "Uniformitarianism," 1–3.
64. Ibid., 2.
65. Ibid. Note that YEC often tries to attach evolutionism to all forms of OEC, which we strongly refute.
66. See Froede, Jr., *Geology By Design*, 20–26.

The Foundations of Correlation

of their ability a classification of rock strata that would account for the phenomena encountered in the crust of the earth.[67]

It essentially comes down to an impasse between those who believe that the Earth is old and those who believe that the Earth is young. Generally speaking, modern science holds to an old Earth and an uniformitarian view of the geologic column. Is it possible that this predominant view among scientists can be reconciled with the Genesis 1–2 record? We believe that it can.

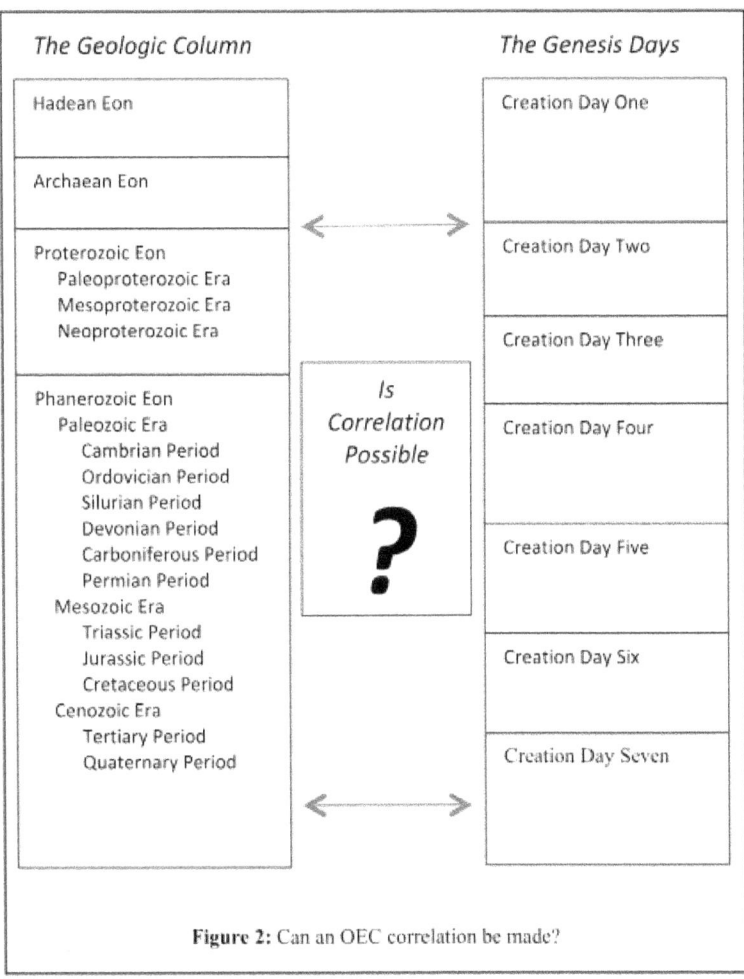

Figure 2: Can an OEC correlation be made?

---

67. Ritland, "Historical Development: Part II," 49.

4

The Correlation Model

THE BASIC PRINCIPLES

HAVING AFFIRMED THAT THE geologic column is more than just "textbook orthodoxy," we now present the correlation model that provides an example of how the modern geologic scale and the Creation Days of Genesis can be found commensurate. This is done in the hope that others will be motivated to do research in this particular area. It is also done in the hope that this model will be improved upon.

The correlation model will be presented as follows. First, since we come with a systematic perspective, the seven Days of Genesis will be aligned with the standard geologic timescale. It is understood that time, as a created entity, existed prior to the Creation Week. Thus, the Creation Week, as presented in Genesis 1–2, is a narrowing of the divine focus from the wider cosmic (Universe) view (Gen 1:1) to the narrower terrestrial (Earth) view (Gen 1:2–2:4a).

Second, since we come with an Old-Earth Progressive Creationism perspective, each "Day" is considered to be a *God Day*. This means that it transcends the typical human conception of a solar day. Each God Day is simply the time that God used to accomplish a given work of creation. This also means that some Days are chronologically longer than other Days. (NOTE: Unless otherwise indicated, the geological dates used in this project are in accordance with the International Geologic Time Scale as ratified by the ICS in 2008.)

Third, since we come with a concordantist perspective, the description given of each Day will freely integrate both biblical revelation and scientific discovery. As a presuppositional reminder, it is understood that the Genesis 1–2 text embodies only the highlights of the divine creative process as it relates to the ultimate divine anthropic purpose (the completion of an inhabitable environment for humanity). Thus, it is understood that scientific discovery will present much more detail about the created order than does the scriptural record.

Fourth, since we come with an "all truth is God's truth" perspective, it is understood that the Genesis text and the geologic column—while each will bring some different truth to the correlation—will not bring contradiction. The postulations of modern science will be framed within the scriptural revelation. Scripture is always understood to be primary. This means that any assertions of science that are considered to be antithetical to scriptural teaching will be evidentially reinterpreted or rejected outright. This includes the concept of evolution.

The Concept of Evolution

Though we have no overt interest for this project in the subject of biological evolution, a few thoughts on the matter are necessary. Despite the widespread acceptance of macroevolution by the wider scientific community, we reject the notion in this model. This rejection is not because we believe that the idea is the spawn of hell, but simply because we do not hold to its veracity.[1]

First, and foremost, we do not believe that the concept of biological macroevolution is compatible with a straightforward understanding of the Holy Scriptures. In fact, it stands in juxtaposition to the "strong and frank creationism" found in the *divine revelation* of Genesis and the greater message in the entirety of the Bible.[2] Divine *ex nihilo* and singularity-based creation is a hallmark of the Hebraic mind as reflected in the scriptural text.[3]

---

1. See Ramm, *Christian View*, 243; and Lennox, *Seven Days*, 167–71. Also, for an excellent and fair discussion, see Newman et al., "The Status of Evolution."

2. Ibid., 58–62. Ramm speaks of "the Bible's strong and frank creationism."

3. See Kohler and Hirsch, "Creation," 336. Here they aver: "Whatever may be the nature of the traditions in Genesis, and however strong may be the presumption that they suggest the existence of an original substance which was reshaped in accordance with the Deity's purposes, it is clear that the Prophets and many of the Psalms accept without reservation the doctrine of creation from nothing by the will of a supermundane

Essentially, by its very nature, evolutionism, grounded in *closed naturalism* with a heavy bent toward random processes and natural selection,[4] remains resistant to the theistic scriptural reality, which, according to the classic canonical teaching of the traditional church, embodies a God-driven *open supernaturalism* with its all-encompassing anthropic metanarrative.[5] In fact, as R. Albert Mohler, Jr. states: "Evolution presents a direct challenge to the entire story-line of the Bible."[6] Deep time is entirely commensurate with the biblical text concerning origins; macroevolutionism stands in stark contrast to the Hebraic-textual impetus.[7]

Second, from a purely empirical perspective, a very strong case can and has been made for the theory of abrupt appearance (contra evolution).[8] It brings to mind the question once asked by geneticist Walter E. Lammerts, "Why not creation?"[9] The fossil record simply shows that the ancient creatures whose remains are found preserved in the strata appear in an orderly progression in their complex form. It does not substantiate that they are morphologically connected to one another or to creatures of today. At the very least, either possibility (creation or evolution) seems to be just as viable as the other.[10] The key to reaching conclusions involves evidentiary

---

personal God (Pss 33:6–9; 102:25; 121:2; Jer 10:12; Isa 42:5; 44: 7–9): 'By the word of the Lord were all the heavens made; and all the host of them by the breath of his mouth.' To such a degree has this found acceptance as the doctrine of the synagogue."

4. Weisz et al., *Science of Biology*, 716–19. They assert the standard academy line: "Living creatures on earth are a direct product of the earth . . . Nothing supernatural appeared to be involved—only time and natural physical and chemical laws operating within the peculiarly suitable earthy environment" (716).

5. Wilson, *Hebraic Heritage*, 21–22. He states: "Scripture emphasizes that theology begins with the supernatural and outward. The revelation of God is mainly perpendicular; it comes through divine initiative. The Bible is not primarily the story of man's search for and discovery of God; rather it is God, in his grace, disclosing himself to man" (22).

6. Mohler, Jr., "Christianity and Evolution," Para. 16.

7. For example, evolutionism posits that macroevolution, of necessity, must be continuing. Contrarily, the Bible posits that God's creative work ceased at the end of the Sixth Day (humanity) issuing in the Seventh Day (Sabbath).

8. Bird, *The Origins of Species Revisited I*, 46–64. He argues that there is strong fossil evidence of abrupt appearance with discontinuity between appearances of natural groups of complex organisms. Bird shows that the paleontological record consists of [1] systematic abrupt appearances of creatures in their complex form with [2] systematic gaps between the appearances. This understanding fits much more cohesively with the scriptural record than does evolution.

9. Lammerts, "Introduction," 1–4.

10. See Klotz, "The Philosophy of Science," 5–23. He said, "To reject evolution is not

The Correlation Model

allowance and interpretive methodology: what should be allowed into the evidence equation and how should it all be *comprehensively* understood in the earnest seeking of truth?[11]

From the view of Christian concordantism, of which the Genesis Column Model is grounded, the primary revelation of Scripture (first-line evidence), which is profoundly creationistic in its presentation, must then be the grid by which the secondary revelation of nature (second-line evidence) is creationistically interpreted.[12] The theory of abrupt appearance, although its postulations are based *entirely* on the natural record, can precisely accomplish this comprehensive coherence with Holy Scripture.

Therefore, for the purposes of this correlation model, we will speak of the advent of biological lifeforms in terms of their God-directed *appearances* rather than any form of evolutionary morphology, whether theistic or otherwise. However, having said this, it must also be noted that the substance of the Genesis Column Model, while being a creationistic and not an evolutionistic paradigm, should not ultimately be affected by an agreement or disagreement with biological evolutionism. The intervals in the model are set by definitive empirical markers and not primarily by a particular view of secondary causative processes.

Third, there is one last related thought. Despite the continual effort of Young-Earth Creationism to assert otherwise (note Sewell), there is a clear distinction between being an advocate of the geologic column with its visible "succession of life-forms" and an advocate of biological macroevolution.[13] The concept of evolution may require the geologic column and deep time, but the geologic column and deep time does not require the concept of evolution. As Eerdman clearly states with reference to her example of the classic Grand Canyon record:

> The Grand Canyon of the Colorado River has often been likened to a book whose pages reveal the story of many eons of earth history. The story is told not alone by the relationship of the rocks to one another and the relationship of the river to the overall structure, but in large measure by the remnants of life preserved

to reject science" (6).

11. See Roth, "Does Evolution Qualify?," 4–10; also, Bird, *Origins Revisited*, 41–104.

12. See Plantinga, "Methodological Naturalism?," 143–54. He argues that "a Christian academic and scientific community ought to pursue science in its own way, *starting from* and taking for granted what we know as Christians" (143).

13. Stearn and Carroll, *Paleontology*, 18

> in these rocks. Here on a vast scale and in an accessible form is a completely unprejudiced account of ancient life.[14]

The rocks and the fossils therein do not cry out for an evolutionary explanation. They are just plainly and visibly there—"a completely unprejudiced account of ancient life." Ultimately, just how this "unprejudiced account" is understood comes down to the variant interpretive factors and perspectival conjectures of researchers.

From an evangelical Christian worldview and in consideration of the Genesis account, we are convinced that the faunal succession so visible in the geologic record is best explained by divine special creation with discontinuity (a biblical variation of abrupt appearance theory) over the course of successive points of deep time. God first created the entirety of his universe (Gen 1:1) and then molded it into its final form through progressive, yet conclusive steps. When new creatures *appear* in the record, it is due to a direct act of completed work by God and not through a macroevolutionary process. There is absolutely no conflict between special creation and deep time. This also means that special creation is not the inherent and sole possession of Young-Earth Creationism. Special creation, as presented in scripture, does not require a young Earth.

EXISTENCE BEFORE CREATION WEEK

Introduction

We begin by presenting a time-line chart illustrating the absolute transition point between pre/space-time (before creation began) and space-time (when/after creation began). This intersection is what is known as the *initial creation singularity*. It is also that point where narrow science can legitimately go no further (the end of nature) in the study of cosmic causation. However, while the work of natural science must necessarily ground to a halt at the *end of nature*, Christian metaphysics, with its ultimate understanding of ontology, indeed can and does go far beyond where science can traverse. The biblical revelation unabashedly traverses beyond finity into the realm of infinity; the Scriptures claim that beyond time is eternity, beyond space is heaven, and beyond everything is the divine First Cause.

---

14. Eerdman, "Fossil Sequence," 1

The Correlation Model

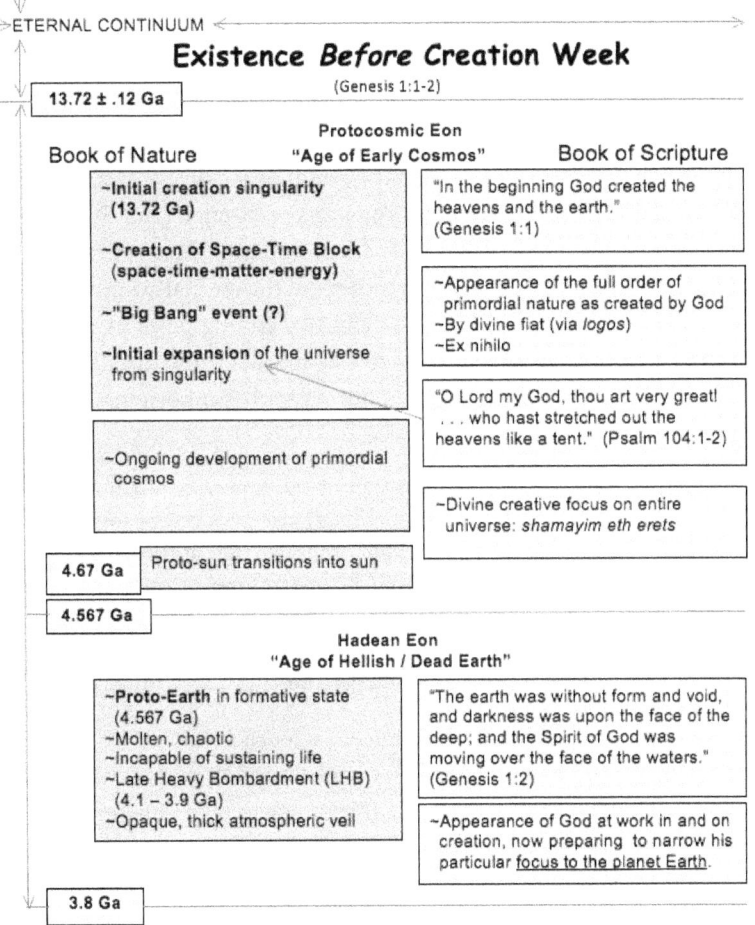

Figure 3: Existence Before Creation Week correlation

Genesis 1:1 boldly proclaims, "In the beginning, God created the heavens and the earth." This is understood to be the beginning (Hebrew *bereshith* = beginning, first[15]) of all creation (Hebrew *shamayim* = heavens, sky, horizon[16]; Hebrew *erets* = whole earth, land, ground, countries[17]). The phrase "the heavens and the earth" (i.e. *shamayim* and *erets* used together)

---

15. Brown et al., *Hebrew and English Lexicon*, "bereshith," 912 (H7225).
16. Ibid., "shamayim," 1029–30 (H8064).
17. Ibid., "erets," 75–76 (H776).

is a common Hebraism for *the universe*.[18] Thus, Genesis 1:1 refers to the precise moment that "the whole creation" (Rom 8:22) came into primordial existence by the mere *fiat* word of its Creator (John 1:1–3; Rom 4:17; Heb 11:3). It was at this point, which modern science currently posits to have been 13.72 billion years ago, that God initially created everything outside of himself and (presumably) outside the realm of the heavenlies. In so doing, God overlaid the spatial-temporal-material existence (i.e., the space-time block) across the pre-existent eternal continuum. Thus, space-time had a definitive beginning point. In the context of the eternal continuum, space-time will be very brief and have a definitive ending point as well. It is the current predominant view of modern science that some variation of the Big Bang Theory is the best explanation for cosmological origination. According to Lambert:

> Earth's origins lie in the creation of the universe. Just how this came about remains unclear, but many scientists accept some version of the big bang theory, which goes like this: at first all energy and matter (then only subatomic particles) were closely concentrated. About 14 billion years ago a vast explosion scattered everything throughout space. Star studies prove that the universe is still expanding, and background radiation hints at its initial heat.[19]

This is an important assertion of science because the Big Bang is singularity-based: namely, there was a precise point of its initiation—a definitive *beginning* (cf. Gen 1:1). Lambert, whether or not thinking of the implications, states that the origin of the Earth lies "in the *creation* of the universe." This implies that the universe was *created* and thus must have a *cause*. This includes, not only particles of matter (and forces of energy), but even space and time.

Universal Causation

Richard Swinburne speaks of two types of causes. There are *scientific causes* (explained by laws and conditions) and *personal causes* (explained by agents and volitions).[20] This brings up a very important question. Is it

---

18. Schrader, "Genesis," 8. Here Schrader elaborates, "It is the whole universe and everything in it that the verse speaks of when it says '. . . heaven and earth. . . .' The Hebrews did not have one word for universe, so they used this expression."

19. Lambert, *Field Guide to Geology*, 16.

20. Swinburne, *Existence of God*, 32–48.

even plausible for a beginning of the universe to have a purely scientific cause? Contrary to certain fanciful postulations advocated by naturalistic science, particularly those from the quantum fields,[21] the clear and commonsensical answer to that question is *no*. Prior to the creation of the space-time universe, there was no space-time universe. Therefore, outside of the mystical spiritual realm, there existed nothing; there was nothing existing in space-time because there was no space-time in existence. It is a commonsensical absurdity to posit that nothing created something out of nothing. Facing this reality is what leads some—who do not hold to a biblical worldview—to propagate extraordinary and extremely absurd theoretical extrapolations that advocate such ideas as eternal space-time (creation has always existed = no need for a creator) and quantum-based ex nihilo non-agency causation (creation created itself = no need for a creator). The explicit bottom-line is this: The motive of closed naturalistic cosmology is to avoid acknowledging an absolute beginning of the universe in order to avoid acknowledging an absolute divine Creator of the universe. Simply put, naturalistic cosmology dodges God!

The Biblical View of Creator Cosmology

What do the Scriptures clearly affirm? The Bible is replete with statements proclaiming the Creator-God and his great attributes. For instance, note the following New Testament texts:

> *John 1:1–3*—"In the beginning was the Word, and the Word was with God, and the Word was God. He was in the beginning with God; all things were made through him, and without him was not anything made that was made."

> *Colossians 1:15–17*—"He is the image of the invisible God, the first-born of all creation; for in him all things were created, in heaven and on earth, visible and invisible, whether thrones or dominions or principalities or authorities—all things were created through him and for him. He is before all things, and in him all things hold together."

---

21. For instance, see Hawking, *Brief History of Time*, 115–41. Hawking has authored several quantum cosmological models that are expressly designed to negate a beginning to the universe and thus deny the need for a Creator of the universe. He reasons: "So long as the universe had a beginning, we could suppose it had a creator. But if the universe is really completely self-contained, having no boundary or edge, it would have neither beginning nor end: it would simply be. What place, then, for a creator? (141)

I Corinthians 8:6—"[Y]et for us there is one God, the Father, from whom are all things and for whom we exist, and one Lord, Jesus Christ, through whom are all things and through whom we exist."

Hebrews 1:1–3—"In many and various ways God spoke of old to our fathers by the prophets; but in these last days he has spoken to us by a Son, whom he appointed the heir of all things, through whom he created the world. He reflects the glory of God and bears the very stamp of his nature, upholding the universe by his word of power."

Hebrews 11:3—"By faith we understand that the world was created by the word of God, so that what is seen was made out of things which do not appear."

The Bible clearly affirms the divine origination of the universe. The scriptures begin (Gen 1:1) with a tri-fold declaration of reality: God exists, creation had a beginning, and God is the Creator of creation. The very fact of creation is evidentially affirming of the existence and action of God. As Romans 1:19–20 confirms:

> For what can be known about God is plain to them, because God has shown it to them. Ever since the creation of the world his invisible nature, namely his eternal power and deity, has been clearly perceived in the things that have been made. So they are without excuse.

Although not complete in and of itself, there is great divine truth available in general revelation. God can indeed be seen and perceived *clearly* in his creation.

The Conceptual View of Creator Cosmology

But—who is he and what are his attributes of Deity as revealed "in the things that have been made"? Craig presents an amazing comprehensive synopsis—a "conceptual analysis"—of the biblical Creator-God from a logical-philosophical perspective as grounded in natural theology:

> If we go the route of postulating some causal agency beyond space and time as being responsible for the origin of the universe, then conceptual analysis enables us to recover a number of striking properties which must be possessed by such an ultra-mundane being. For as the cause of space and time, this entity must transcend

THE CORRELATION MODEL

> space and time and therefore exist atemporally and non-spatially, at least sans the universe. This transcendent cause must therefore be changeless and immaterial, since timelessness entails changelessness, and changelessness implies immateriality. Such a cause must be beginningless and uncaused, at least in the sense of lacking any antecedent causal conditions. Ockham's Razor will shave away further causes, since we should not multiply causes beyond necessity. This entity must be unimaginably powerful, since it created the universe without any material causes. Finally, and most remarkably, such a transcendent cause is plausibly taken to be personal.[22]

He elaborates further concerning this *personal* transcendent cause (i.e., personal Creator-God):

> Moreover the personhood of the cause of the universe is implied by its timelessness and immateriality, since the only entities we know of which can possess such properties are either minds or abstract objects, and abstract objects do not stand in causal relations. Therefore, the transcendent cause of the origin of the universe must be of the order of the mind. This same conclusion is also implied by the fact that we have in this case the origin of a temporal effect from a timeless cause. If the cause of the origin of the universe were an impersonal set of necessary and sufficient conditions, it would be impossible for the cause to exist without its effect. For if the necessary and sufficient conditions of the effect are timelessly given, then their effect must be given as well. The only way for the cause to be timeless and changeless but for its effects to originate *de novo* a finite time ago is for the cause to be a personal agent who freely chooses to bring about an effect without antecedent determining conditions. Thus, we are brought, not merely to a transcendent cause of the universe, but to its personal creator.[23]

Since the cause of the space-time universe must be timeless and non-spatial, the cause of the universe must be either of the nature of mind (personal) or of the nature of abstract object (impersonal). Since abstract objects are just that—abstract—they cannot have causative effect. Thus, the cause must then be one who is possessing of mind. Furthermore, the cause of the universe cannot be merely an impersonal set of necessary and sufficient conditions, because then the cause (creator) and its effects (creation) would

22. Craig, "Ultimate Question of Origins," 10.
23. Ibid., 10–11.

be one and the same and thus inseparable (pantheism). Therefore, in conclusion, the transcendent cause of the universe must be a thought-capable person with a freedom of choice and who is existent apart from the creative effects of the cause. This certainly affirms the Creator-God cosmology of the biblical revelation.

The Wider Pre-Creation Week Cosmology: The Universe

This brings us back to the creative work of God prior to the Creation Week as it is presented in Scripture. Following the ex nihilo absolute singularity (Gen. 1:1; cf. Heb 11:3), God began his process of "making" (Hebrew *asah* = crafting) the universe (Ps 104:2—he "stretched out the heavens like a tent"—cosmic expansion?) that had already come into material existence (Gen 1:1—Hebrew *bara* = created) through the mere power of his Word (John 1—Greek *logos*). Like a potter molding clay, God crafted stars and planets and all the details of the greater universe. They are all his *personal* "handiwork" (Ps 19:1). The entirety of creation bears his fingerprint (Ps 8:3). Science holds that some nine billion years passed in chronological time between the initial singularity (13.72 Ga) and the forming of the proto-Earth (4.56 Ga). In our model, this almost incomprehensible period of time will be called the *Protocosmic Eon* (lit., "time of primordial universe"). Despite its enormous span of deep time, this eon—from the biblical perspective—is always to be understood relative to the Earth. For, in this understanding, as soon as the primeval proto-Earth initially formed, a new eon (the Hadean) began. This means that God, from the human temporal perspective,[24] spent epochs of time personally fine-tuning the intricacies of the greater cosmos for his ultimate anthropic purpose. This is a purpose played out with the Earth as the center stage and humanity as the primary and central feature. The Protocosmic Eon (Gen 1:1) was essentially the time that God took to make everything ready for the specific planet that would be the habitation place of humanity. Then, and only then, did God turn his creative focus upon the final environment: the young proto-Earth (Gen 1:2).

---

24. See Ps 90:4 (cf. 2 Pet 3:8) for a God-human comparative concerning chronological time.

The Narrowing of God's Creative Focus: The Earth

Genesis 1:2 states: "The earth was without form and void, and darkness was upon the face of the deep; and the Spirit of God was moving over the face of the waters." The words used in this passage are extremely significant: "without form" (Hebrew *tohuw* = formless, desolate, confused[25]), "void" (Hebrew *bohuw* = empty[26]), "darkness" (Hebrew *choshek* = extraordinary darkness, obscure, distressed, miserable[27]), "the deep" (Hebrew *tehom* = deep, sea, abyss[28]), and "the waters" (Hebrew *mayim* = waters, usually moving; waters of danger and violence[29]). When viewing the scene presented in totality, we suggest that the conditions described are a primordial undeveloped existence that was chaotic (*tohuw*), empty (*bohuw*), covered in darkness (*choshek*), and an abyss (*tehom*) of constant violent upheaval (*mayim*). Using very graphic phenomenological imagery, the text amazingly describes the conditions of the developing proto-Earth. According to science, the conditions of Earth at this stage were "hellish," thus the name Hadean Eon (i.e., derived from *Hades*, lit., "the abode of the dead"[30]). At that time, the primeval Earth was a chaotic, yet dead, existence—a volatile environment completely incapable of supporting life.

During the earliest part of the Hadean Eon, the Solar System began to form by coagulation and gravitational attraction from the accretion disk of the early proto-Sun. The proto-Sun was surrounded by a massive and thick gas and dust cloud. At some point, the sun began to collapse in on itself due to gravitational compaction. It eventually reached a stage of collapse where the proto-Sun ignited through nuclear fusion and began to give off heat and light. Meanwhile, the particles in the cloud surrounding the young Sun began to coalesce into planetesimals. Eventually the planetesimals began to aggregate together to form microplanets. The microplanets and planetesimals engaged in much collision activity. The combination of the heat generated through these massive collisions with the innate gravitational heating and the radiation present within the interior of the bodies

---

25. Brown et al., *Hebrew and English Lexicon*, "tohuw," 1062 (H8414).
26. Ibid., "bohuw," 96 (H922).
27. Ibid., "choshek," 365 (H2822).
28. Ibid., "tehom," 1062 (H8415).
29. Ibid., "mayim," 565–66 (H4325).
30. Mounce, *Expository Dictionary*, "Hades," 331 (H86). Although different in Hebraic thought, *Hades* is often associated with *Hell*.

produced incredible temperatures. Therefore the proto-Earth and the other proto-planets of the Solar System were initially molten.[31]

The Earth's Moon formed much later than the early accretion activity. As posited by Ogg, it is thought that a Mars-sized body (Theia) collided with the proto-Earth during this period. In this theorized collision, a molten Theia was absorbed into the Earth's molten body adding an additional ten percent mass. The formation of the Moon occurred from the gravitational coalescence of the orbiting molten debris stemming from the collision.[32]

The primeval history of the Solar System was extraordinarily chaotic. As the molten proto-planets, including Earth, coalesced, their gravitational pull continued to attract other smaller bodies that were still orbital. This resulted in what is known as Late Heavy Bombardment (LHB). These bodies relentlessly showered the Earth and Moon. (The scars from LHB can still be seen on the lunar surface in the form of craters.) Further, the terrestrial crust was developing. This also was an extremely turbulent process as the crust would continuously form and be reformed over the course of millions of years. Interestingly, unlike the Moon, which has very little internal movement, it was this very process of crustal forming-reforming that virtually eliminated any LHB scars from the Earth's surface. M. Alan Kazlev and Franco Maria Boschetto describe this process:

> This is the period in which the Earth's crust was formed. This crust melted and reformed numerous times, because it was continuously broken up by gigantic magma currents that erupted from the depths of the planet, tore the thin crust, and then cooled off on the surface before sinking again into the heart of the Earth. The details of this slow, destructive process are still uncertain. However, it is thought that the heavy elements, like iron, tended to sink toward the center of the Earth because of their higher density, while the lighter components, particularly the silicates, formed an incandescent ocean of melted rock on the surface. Approximately 500 million years after the birth of the Earth, this incandescent landscape began to cool off. When the temperature fell under 1000° C., the regions of lower temperatures consolidated, became more stable, and initiated the assembly of the future crust.[33]

They continue with this thought:

---

31. Kazlev and Boschetto, "Hadean Eon," 2.
32. Ibid.
33. Ibid., 3.

The Correlation Model

> In principle therefore the Earth was a sphere of melted rock, churned by convective movements between the hot inner layers, while the outer, surface regions were in contact with the cold of surrounding space. The dissipation of heat to space began the cooling of our planet. In the magma ocean, blocks began to appear, formed from high melting point minerals. These red hot, but solid slabs were similar (although on a very different scale) to the thin edges of crust that we see forming on the surface of flowing lava ... Truly a nightmare landscape![34]

The Hadean Eon is not considered to be a strict geologic age. This is because there are no known rocks on Earth from this period. The proto-Earth did not reach the point of becoming solid rock until approximately 3.8 billion years ago.[35] Thus, a large portion of Hadean time represents the molten and grossly turbulent period of geohistory. According to Ben M. Waggoner:

> Because collisions between large planetesimals release a lot of heat, the Earth and other planets would have been molten at the beginning of their histories. Solidification of the molten material into rocks happened as the Earth cooled. The oldest meteorites and lunar rocks are about 4.5 billion years old, but the oldest Earth rocks currently known are 3.8 billion years old. Sometime during the first 800 million or so years of its history, the surface of the Earth changed from liquid to solid. Once solid rock formed on the Earth, its geological history began.[36]

It is also important to understand other dynamics of the early formation of the Earth. The three hallmarks of the Hadean Eon-Archean Eon transition (3.8 Ga) are [1] the solidification of the Earth, [2] the beginning of a true atmosphere, and [3] the development of a functioning water cycle. Due to peak volcanic activity and to the incredible temperatures at the center of the Earth, the crust was emitting massive amounts of thick and toxic gases. These included methane, halogen gases, nitrogen, carbon dioxide, hydrogen, ammonia, water vapor, and many others. Over a period of at least 100 million years, these gases accumulated to form the Earth's early noxious atmosphere. The blanket of gases provided for an extremely dense and opaque atmospheric cover. It is thought that the primeval atmosphere

---

34. Ibid.
35. Waggoner, "Hadean time: 4.5 to 3.8 billion years ago."
36. Ibid.

was very similar to that of Titan, one of Saturn's large moons. Incredibly, it is also thought that this early atmosphere had about 250 times the current level of atmospheric pressure![37]

During the long epoch of Hadean time, an amazing transition began to occur. Not only did the outgasing of the Earth begin building an atmosphere, but as a result, a water cycle began to be established as well. Kazlev and Boschetto explain:

> In fact, the primitive Earth long remained covered in darkness, wrapped in dense burning clouds into which continuously poured water vapor from volcanic emissions. When temperatures finally cooled sufficiently, the clouds began to melt into rain, and the primordial atmosphere began to produce storms of unimaginable proportions, under which the Earth groaned and flowed. At first, falling on incandescent rock, the rain evaporated, but the evaporation gradually cooled the crust until the water could accumulate in the depressed regions of the Earth's surface, forming the first oceans.[38]

This process eventually led to the formation of sedimentary rock. With the advent of the early water cycle and the cooling of surficial conditions, such processes as erosion, sediment accumulation, and drift began. As erosion occurred to elevated surfaces, the detritus (erosive sediments) became deposited to the bottom of the primordial oceans where it eventually worked its way back into the molten realm. In time, due to the remelting of these erosive sediments, magmas were produced that were rich in silicates. With continuation of the crustal recycling, this ultimately resulted in the formation of a new lower density, granitic crust that began to replace the early higher density, basaltic crust. This is significant because the granite crust both floats on basalt and endures incredible forces such as metamorphics and tectonics without breaking up or without sinking back into the depths of the Earth.[39] Therefore, as the Hadean Eon began to conclude, the planet—though still primitive—had undergone a startling metamorphosis. Earth had transitioned from a molten proto-planet into a terrestrial planet with three key components: a dense gas zone (atmosphere), a liquid zone with a functioning water cycle (hydrosphere), and a solid zone with the

---

37. Kazlev and Boschetto, "Hadean Eon," 5.
38. Ibid., 6.
39. Ibid., 6–7.

beginnings of a modern crust that would eventually develop continental land masses (lithosphere).⁴⁰

The Two Key Images of Genesis 1:2

In Genesis 1:2, there are two key images present. First, there is the image of the primordial Earth in its extreme and chaotic state. Second, there is the image of the "Spirit of God" (Hebrew *ruwach elohiym*: *ruwach* = breath, wind, spirit[41]; *elohiym* = God, divine plurality[42]) "hovering" (Hebrew *rachaph* = hover[43]) "over" (Hebrew *al* = upon, over, above[44]) the "presence" (Hebrew *paniym* = face, presence, sight[45]) of the violent and molten abyss. This is important. As Theodore H. Robinson says of the proto-Earth, "There is no order, no shape, no distinction between solid and liquid, nothing but a confused indiscriminate mixture, and, hovering above it as a bird over its nest, the Breath of God, already, it seems, personified."[46] Notice Robinson's description of *ruwach elohim* as the *Breath* of God. It is quite appropriate to the context, for the Spirit of God would in time breathe life into the dead created order (cf. Gen 1:11–12; 20–21; 24–25; 26–27; esp. 1:30 and 2:7). At the appropriate point of divine *kairos*, God would indeed cause organic life to appear. The Holy Spirit had entered the realm of the "hellish" proto-Earth and begun to personally craft a devastatingly incomplete and chaotic dead state into a state of increasing order that would eventually support life and be proclaimed by God to be "very good." The work of God in Genesis 1:2 set the stage for the beginning of the Creation Week. In other words, God's work during the Hadean Eon served as both a preparation and a transition for the incredible progression of divine creative events that he would initiate and accomplish over the next *God Days*.[47] God entirely

---

40. Ibid., 5.
41. Brown et al., *Hebrew and English Lexicon*, "ruwach," 924–25 (H7307).
42. Ibid., "elohiym," 43–44 (H430).
43. Ibid., "rachaph," 934 (H7363).
44. Ibid., "al," 1106 (H5921).
45. Ibid., "paniym," 815–19 (H6440).
46. Robinson, "Genesis," 220.
47. Since God exists in *kairos* (purpose time—divine perspective) and is not bound by *chronos* (clock time—human perspective) as we are, perhaps it only seems to us that he used time as we understand it. God is God; he creates and does all that he does in the fullness of his time and in the manner and for the purpose that he wills.

pursued his creative work with the advent of life, and ultimately human life, foremost as his purpose.

GOD DAY ONE

Introduction

This begins the Creation Week of Genesis. In our model, God Day One and the Archean Eon (3.8 Ga–2.5 Ga) are considered to be rough equivalents. This period lasted about 1.3 billion years. It is the epoch when Earth's true geologic time began.

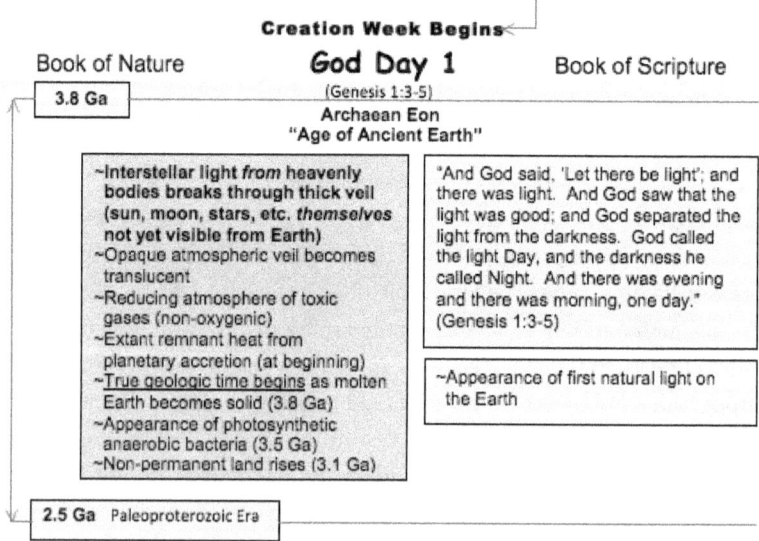

Figure 4: God Day One correlation

The Scriptural Text: God Day One

Genesis 1:3–5 states: "And God said, 'Let there be light'; and there was light. AndGod saw that the light was good; and God separated the light from the darkness. God called the light Day, and the darkness he called Night. And there was evening and there was morning, one day."

The Significance of God Day One

On this first Day, God caused a significant event to occur. The Earth, though still in extremely primitive conditions and vastly different from the planetary conditions of today,[48] had its dense, opaque atmosphere of darkness shattered by the first waves of interstellar *light* (Hebrew *owr* = light, light as diffused in nature, morning light, light of heavenly luminaries, daylight, lightning[49]) from the Sun and from distant stars. Sailhamer puts the text in perspective:

> Verse 3 has often been taken to mean that God created light before he had created the sun, since here he said, "Let there be light," but not until v. 16 does the narrative speak of God making the sun. It should be noted, however, that the sun, moon, and stars are all to be included in the usual meaning of the phrase "heavens and the earth," and thus according to the present account these celestial bodies were all created in v. 1. Verse 3 then does not describe the creation of the sun but the appearance of the sun through the darkness.[50]

From a scientific viewpoint, several developing conditions led to this change. Around 3.8 billion years ago, though there still existed residual heat from the initial planetary accretion, the molten crust of the proto-Earth had begun to harden into a solid state. As more and more rock formed, the first continental plates began to rise over the seas of magma and the crustal conditions began to stabilize. With the increasing solidification of the crust and an increasing surficial equilibrium, the noxious outgassing from the recesses of the Earth began to decrease over time. Combined with the massive rainstorms of the Hadean-Archean transition that cooled the surface and enhanced the development of the granitic crust through the formation of oceans, the atmospheric and surficial conditions slowly became more conducive to the advent of primitive life.[51] Though still extremely toxic to most life on Earth today, the dense thick opaque cloud began to thin and allow for some light to diffuse through the atmosphere to the surface. In

---

48. Speer, "Introduction to the Archean."

49. Brown et al., *Hebrew and English Lexicon*, "owr," 21 (H216).

50. Sailhamer, "Genesis," 26. We would modify and say "the appearance of the *light* (not yet the sun) through the darkness."

51. For a very influential paper on how the outgassing of the proto-Earth effected the development of the early atmosphere and hydrosphere, see Rubey, "Geologic History of Sea Water," 1111–48.

effect, the Earth's atmosphere began to change from an opaque veil to a translucent veil. As the darkness increasingly dispelled, more and more light came through. This diffusion of light provided for the beginnings of photosynthesis. The result was the advent of first life. As Brian R. Speer says:

> It was early in the Archean that life first appeared on Earth. Our oldest fossils date to roughly 3.5 billion years ago, and consist of bacteria microfossils. In fact, all life during the more than one billion years of the Archean was bacterial.[52]

Kevin Hefferan elaborates:

> The Archean is known as the "Age of Prokaryotes," the first documented forms of life. Prokaryotes are simple, single-celled organisms that lack a nucleus, mitochondria and organized cell function. Prokaryotes have a distinct advantage over other life forms in that they reproduce asexually . . . asexual reproduction precludes new genetic input, resulting in the same old, same old. However, this type of reproduction allows for the duplication of great numbers of individuals in a cookie cutter fashion. Therein lies the advantage: quick and simple reproduction. Prokaryotes include bacteria and—most important in our story—cyanobacteria![53]

Hefferan continues: "Why are cyanobacteria important? . . . Through photosynthetic activity, cyanobacteria fundamentally altered the Archean atmosphere through the production of a valuable waste product—free oxygen!"[54] It is important to note that the free oxygen produced during the Archean Eon was largely absorbed into the Earth by the developing rocks and thus was not at that time inundating the atmosphere itself. Therefore, the life forms during the Archean were photosynthetic anaerobes. They expelled oxygen as waste, but did not breath oxygen As more and more cyanobacteria rapidly reproduced asexually, the more free oxygen was expelled into the Earth's environment. It would not be until the Earth's lithosphere and hydrosphere reached full oxygenic saturation that the atmosphere would begin to become oxygenated.[55]

---

52. Speer, "Introduction to the Archean."
53. Hefferan, "Archean Life," 1.
54. Ibid., 1–2. Note: Oxygenation of the atmosphere would later occur substantially during the ensuing Proterozoic Eon.
55. Lane, "First Breath," 36–39.

THE CORRELATION MODEL

Some key contributions to the development of the Earth that occurred during the Archean Eon are [1] the entrance of significant light to the planet coinciding with the decreasing of the thick atmospheric veil (opaque to translucent) providing for the beginning of photosynthesis; [2] the development of first life (anaerobic) that provided for the early oxygenation of the environment; and [3] the formation of the majority of continental masses (whose basement complexes are called *shields*[56]) through volcanic outpourings below the primeval ocean. It should also be noted that large portions of the Earth, quite possibly the entire planet, was covered with ocean at this time.[57] It is believed that at least 70 percent of the continental masses were formed from 3.0 to 2.5 billion years ago. These masses, while remaining sub-oceanic during the Archean, became the early foundations for the continents of the present day.[58]

GOD DAY TWO

Introduction

In our model, the Proterozoic Eon (2.5 Ga–542 Ma; i.e., the age of "First Life") begins God Day Two. We also have included the Cambrian Period (542 Ma–485.4 Ma), the first period of the Phanerozoic Eon (i.e., the age of "Visible Life"), as the concluding part of this Day. The key event during this period was the formation of an atmosphere conducive to aerobic life (namely, organisms that breathe oxygen). The toxic reducing atmosphere (methane, ammonia, etc.) dominant in the Archean Eon, slowly at first—and then very abruptly, changed into an oxygenated atmosphere during the Proterozoic Eon. This development led to the appearance of more complex forms of life.

---

56. Bates and Jackson, *Dictionary of Geological Terms*, 463. Shields are the continental nuclei and are made up entirely of Precambrian crystalline rock (basement rock).

57. Kazlev, "The Archean," 1.

58. Ibid., 1–2.

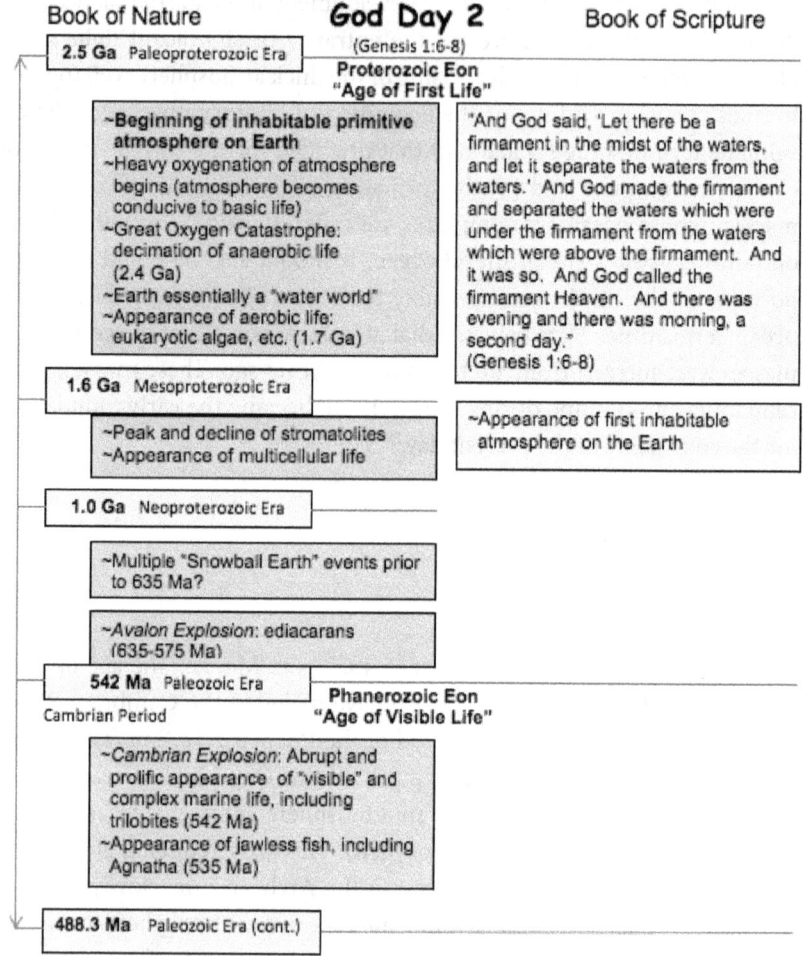

Figure 5: God Day Two correlation

The Scriptural Text: God Day Two

Genesis 1:6–8 states: "And God said, 'Let there be a firmament in the midst of the waters, and let it separate the waters from the waters.' And God made the firmament and separated the waters which were under the firmament from the waters which were above the firmament. And it was so. And God called the firmament Heaven. And there was evening and there was morning, a second day."

The Significance of God Day Two

It is very striking that in this scriptural text, the image of "waters" is used so prominently. However, just as it was in the later phase of the Archean, the Proterozoic Earth was primarily a "water world." Initially, it also retained the Archean reducing atmosphere. Genesis 1:6–8 presents an excellent phenomenological description of the formation of a proto-modern atmosphere amidst an Earth covered by ocean.

According to modern science, the many millions of years of oxygen production by oceanic cyanobacteria ("blue-green algae") ultimately caused the lithosphere/hydrosphere to reach maximum oxygen saturation. This probably occurred between 2.5–2.4 billion years ago. Once the Earth maximally saturated, the continued production of oxygen was forced upward into the atmosphere. Over the course of time, the atmosphere became strongly inundated with oxygen. The *Great Oxygen Event* (GOE) is traditionally dated around 2.4 billion years ago. It is also often called the "Great Oxygen Catastrophe" because it was catastrophic for the majority of extant life forms at that time (which were anaerobic). In effect, the anaerobes almost caused their own extinction.[59] The fossil record shows a sudden end to the vast majority of anaerobic prokaryotes at the 2.4 Ga boundary (although they have continued to exist—even to this day—in isolated anoxic pockets). The GOE, on the other hand, was a blessed event for modern aerobic life. The oxygenation of the atmosphere made the advanced life of today possible, beginning between 2.2 and 1.7 billion years ago with simple eukaryotic life (aerobic organisms).[60]

Later, there were at least two mass explosions of aerobic life. The first, the Avalon Explosion (ediacarans—complex macroscopic life forms), occurred near the conclusion of the Proterozoic Eon at 575 Ma (perhaps stirring as early at 635 Ma). Bing Shen et al. comments:

> Ediacara fossils, 575 to 542 million years ago, represent Earth's oldest known complex macroscopic life forms, but their morphological history is poorly understood . . . The Avalon morphospace expansion mirrors the Cambrian explosion, and both events may reflect similar underlying mechanisms. The evolutionary history

---

59. White, "Proterozoic," 1. The GOE may also have triggered multiple Proterozoic global glaciation events known as "Snowball Earth" (See Ogg et al., 31–32).

60. Woese and Gogarten, "When did eukaryotic cells first evolve?," 2.

of macroscopic organisms in the late Ediacaran Period (circa 575 to 542 Ma) is regarded as a prelude to the Cambrian explosion.[61]

The second—the Cambrian Explosion (metazoans—complex visible marine life)—was the transitional event initiating the Phanerozoic Eon about 542 million years ago. L. James Gibson comments:

> Perhaps the most compelling feature of the fossil record is the sudden appearance of a wide diversity of fossils at and near the base of the Phanerozoic sediments. This sudden appearance is called the Cambrian Explosion, and has been the subject of much comment and analysis. First appearances of phyla and classes of metazoans (multicellular animals) are not distributed evenly throughout the geologic column, but are largely clustered at the lower end of the Phanerozoic, predominantly from the uppermost Precambrian to the Ordovician, peaking in the Cambrian.[62]

These life events, both so clear and *abrupt* in the fossil record, can—we believe—best be explained as acts of direct special creation. Though evolutionism tries hard to explain these events developmentally and monophyletically, there is no cogent and satisfactory answer within that system.[63] In the fossil record, they are both abrupt and massive *appearances* of new life (thus, their common descriptions as *explosions*). These life forms did not exist before and then, without any prior signs or warning (stratigraphically speaking), they existed in extraordinarily massive and ordered polyphyletic populations. Gibson asserts:

> According to standard evolutionary theory, all organisms derive from a single ancestral species. Darwin's famous book is noted for having only one illustration—the familiar monophyletic evolutionary tree, showing all living organisms linked to a single ancestor.... Unfortunately for the theory, this description is the opposite of the actual pattern in the rocks.[64]

---

61. Shen et al., "Avalon Explosion: Evolution of Ediacara Morphospace," 81–84.

62. Gibson, "Polyphyly and the Cambrian Explosion," 3–6.

63. For further, see Dehaan and Wiester, "Cambrian Explosion," 145–56. Note especially 148–51. Evolutionism is shown to be an inadequate explanation. For a contrasting view, see Scott, *Evolution Vs. Creationism*, 169–74. She attempts to navigate the issue with this statement, "Paleontologists consider the Cambrian Explosion to be an interesting scientific puzzle, but by no means a 'problem' for evolution or even evolution by natural selection" (170).

64. Ibid.

The Correlation Model

He continues:

> Several phyla are soft-bodied and/or microscopic, and absent or very rare in fossils . . . No one would claim the fossil record is perfectly complete, but it does not seem to be bad enough to explain the Cambrian Explosion in terms of monophyly. Fossils of soft-bodied organisms are famously found in Cambrian Preservat-Lagerstatten such as the Burgess Shale and the Chengjiang locality in China. Fossil bacteria are reported from both Precambrian and Phanerozoic rocks. Why would depositional conditions favor preservation of bacteria in both Precambrian and Phanerozoic rocks, but soft-bodied multicellular organisms only in the Phanerozoic and uppermost Precambrian? The fossil record is obviously incomplete, but there is no evidence it is so incomplete it would not preserve fossils of soft-bodied organisms for half their supposed geologic history.[65]

The point is that the geologic record, while not entirely complete in all locations, is indeed accurate in its extant composite presentations, particularly in the abruptness of the Avalon and Cambrian events. Direct creation is a very reasonable explanation for both events—and for all of life. With Gibson (and Ramm), it is our contention that the appearance of any new life is the result of intelligently directed vertical creation (namely, top-down and direct by God).[66]

Another very significant development that occurred during the Proterozoic was the beginning of modern plate tectonics. According to A. Toby White:

> Although continents were small, they consisted of stable cratons. Mid-ocean spreading ridges did a good deal of the moving, just as they do today. However, everything happened a good deal faster. The magma on which the continents floated was hotter, less viscous, and closer to the surface. Hot spots were probably hotter. The continents moved swiftly, collided more often and tended to fracture or suture with greater frequency.[67]

This would ultimately lead to the rise of true and inhabitable landmasses and, in our correlation model, form the boundary between this Day and God Day Three.

65. Ibid.
66. Ramm, *Christian View*, 191.
67. White, "Proterozoic," 1.

GOD DAY THREE

Introduction

Among evangelicals, the matter of the first appearance of land on Earth, based on the postulation of modern science and how it fits in with the Genesis Creation Week, has been a controversial issue. Science claims that the first land may have risen above the ancient ocean during the Archean Eon at around 3.1 billion years ago. However, there is some significant degree of speculation involved in drawing definitive tectonic conclusions going back that far into deep time. Science claims to comfortably extrapolate back to the early supercontinent of Rodinia (1.1 Ga–760 Ma) using the patterns of modern tectonics, but beyond that it is uncertain.[68] Still, all of the scientific postulations about first land seem to be in chronological conflict with the biblical schematic. In our Old-Earth paradigm, a relative dating of land at 3.1 billion years ago places its appearance into Day One. The Scriptures place the appearance of dry land in Day Three. Does this mean that, in an OEC paradigm, the ordering of the Genesis text and the scientific interpretation of natural revelation are not reconcilable concerning this particular matter? We do not believe that to be the case and thus offer a plausible solution.

---

68. Chen et al., "Geophysical Detection of Relict Metasomatism," 1089–91. They suggest that plate tectonics were operating as early as 3.5 Ga.

The Correlation Model

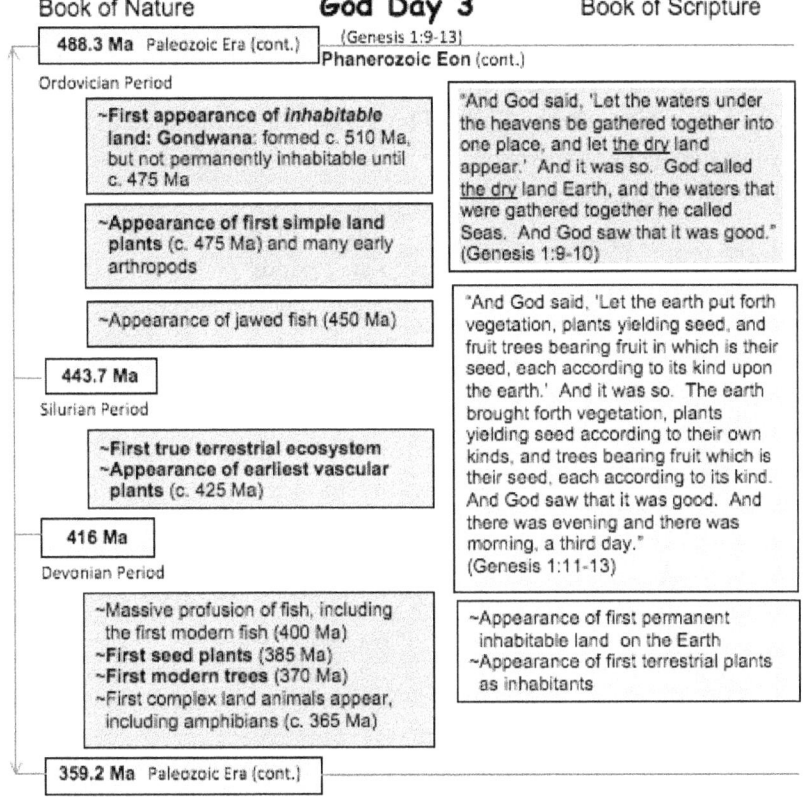

Figure 6: God Day Three correlation

The Scriptural Text: God Day Three

Genesis 1:9–13 states: "And God said, 'Let the waters under the heavens be gathered together into one place, and let the dry land appear.' And it was so. God called the dry land Earth, and the waters that were gathered together he called Seas. And God saw that it was good. And God said, 'Let the earth put forth vegetation, plants yielding seed, and fruit trees bearing fruit in which is their seed, each according to its kind upon the earth.' And it was so. The earth brought forth vegetation, plants yielding seed according to their own kinds, and trees bearing fruit in which is their seed, each according to its kind. And God saw that it was good. And there was evening and there was morning, a third day."

The Significance of God Day Three

The biblical revelation presents two major events on Day Three. First, there is the appearance of some form of dry land. Second, there is the appearance of the first land vegetation. Both of these events are connected in that they involve the divine preparation of land for the advent of advanced life and for the ultimate good of humanity.[69]

In our model, we have equated the Ordovician (488.3–443.7 Ma), Silurian (443.7–416 Ma), and Devonian Periods (416–359.2 Ma) with God Day Three. Each of these periods embodies important occurrences of natural history that are tied to the preparation of the land. Genesis 1:11 provides three descriptions of plants as God brought life into being on the dry land. First, there is a general term used: "vegetation" (Hebrew *deshe* = fresh shoots, grass[70]). Additionally, there are two more specific terms used: "seed-bearing plants" (Hebrew *eseb zara zera*: *eseb* = plants, herb, herbage[71]; *zara* = yielding, sow, scatter[72]; *zera* = seed, offspring[73]) and "fruit trees" (Hebrew *ets periy*: *ets* = trees, wood, woody flax on a stalk[74]; *periy* = fruit[75]). The usage of these particular words is significant for three reasons. First, by use of the general term (vegetation), it means that God instituted a lasting provision of terrestrial plant life that would continue to grow abundantly; there would be a "greening" of the land. Second, by use of the more specific terms (seed-bearing plants and fruit trees), God is not referencing taxonomic narrowness, but rather using phenomenological imagery to reinforce the notion of abundance; the key is the words "seed" and "fruit"—they imply growth (seeding) and profusion (fruitfulness). Very significantly, Genesis 1:12 follows with this result: "The earth brought forth vegetation, plants yielding seed according to their own kinds, and trees bearing fruit in which is their seed, each according to its own kind." This is essentially the application to the plant kingdom of the divine imperative to "Be *fruitful* (fruit) and *multiply* (seed)"—that is, the worldwide expansion of plant life. It sets the stage for the creation of the animal kingdom and for the ultimate ap-

---

69. Sailhamer, "Genesis," 31.
70. Brown et al., *Hebrew and English Lexicon*, "deshe," 206 (H1877).
71. Ibid., "eseb," 793 (H6212).
72. Ibid., "zara," 281–82 (H2232).
73. Ibid., "zera," 282–83 (H2233).
74. Ibid., "ets," 781–82 (H6086).
75. Ibid., "periy," 826 (H6529).

The Correlation Model

pearance of humanity. Third, the images of seeds and fruit stand in stark contrast with the prior bleak environment; the text stresses that God was *emphatically* doing a new thing. In fact, this fits the thoughts of Hans Steur, a paleobotanist, who claims that the formation of land vegetation was "the beginning of an amazing development, which created the terms for animal life on land."[76] The Genesis text paints a phenomenological picture of both supercontinent formation (the advent of dry land—an inhabitable environment for modern life) and purposeful land development (the advent of land plants—terrestrial life itself—the first true terrestrial ecosystem). The appearance of the plants are God's confirming word that the land is now permanent, stable, and ready for new life to begin.

The Advent of Dry Land

According to science, a supercontinent is a large landmass that consists of multiple lithospheric plates (continental cores) that have collided and fused together. Over the course of millions of years, the supercontinents (with plates in constant lateral movement) become unstable because they act as thermal lids limiting the escape of the internal heat of the Earth. Their breakup into smaller continental fragments occurs when the suture points separate due to the continuous thermal pressure.[77] This supercontinent cycle (namely, supercontinent formation, breakup, and dispersal in systematic repeats) has continued and resulted in different landmass amalgamations throughout the history of the Earth.[78]

The succession of supercontinents is thought to be somewhat as follows: Vaalbara (3.1–2.8 Ga), Ur (3.0–1.5); Kenorland (2.7–2.1 Ga); Columbia (1.8–1.5 Ga); Rodinia (1.1 Ga–750 Ma); Pannotia (600–550 Ma); Pangaea (300–180 Ma); Laurasia, Gondwanaland; these were followed by the present day continental formations.[79] Though many details of land formation prior to Rodinia are not definitive, it is known that there have been many instances of landmasses forming and reforming in accordance with tectonic processes going "back into the almost indecipherable past."[80]

---

76. Steur, "Fossil Plants."

77. Lerner and Lerner, "Supercontinents," 573.

78. Ibid. See also Murphy and Nance, "Earth Sciences: On the Assembly of Supercontinents," 324.

79. Nance et al., "Supercontinent Cycle," 72–79.

80. Nield, *Supercontinent*, xiii.

However, beginning with Rodinia and since, science seems to have a good understanding of the successions. Ted Nield calls it "a stately quadrille . . . the grandest of all the patterns in nature."[81] He explains the important connection of the current continents with Pangaea (lit., "all land"), which was the last single planetary landmass prior to this present day:

> The continents of today's Earth are the wreckage of that single supercontinent, Pangaea, which began to break up about 250 million years ago . . . Pangaea consisted of two smaller supercontinents joined at the hip in the region of the Equator: Laurasia in the Northern Hemisphere (North America, Greenland, Europe and much of what is now Asia) and Gondwanaland in the Southern Hemisphere, comprising South America, Africa, India, Australia, and Antarctica. The world we see today is no more than Pangaea's smashed remains, the fragments of the dinner plate that dropped on the floor.[82]

A single common denominator throughout the supercontinent cycle from the Rodinian breakup through the breakup of Pangaea was the supercontinent of Gondwana (650–130 Ma). Kazlev speaks of the significance of Gondwana in Earth's natural history:

> Science tells us that the continents of Australia, India, South America, Africa, and Antarctica, existed together as a separate landmass as long as 650 million years ago. And as these continents only began to break up some 130 million years ago, this great supercontinent had a life of around 520 million years; making it perhaps the most important geological structure of the last billion years.[83]

It is very probable that it was on one of the incarnations of Gondwana that the first land life appeared. This seems to be a favorable possibility due to both Gondwana's long duration (520 million years) and to its consistent equatorial to hemispherically southern location. Moreover, present day Oman (where the oldest plant fossils have been found—see Steur below), located on the Arabian Peninsula, was a part of East Gondwana.[84]

---

81. Ibid.

82. Ibid., 15.

83. Kazlev, "Gondwana."

84. Hauser et al., "Break-up of East Gondwana," 145–57. See also Wopfner, "Malagasy Rift, a chasm in the Tethyan margin of Gondwana," 451–81.

The Advent of Land Life

It is important to realize that the early continents were uninhabitable until the middle Ordovician Period. It was during this general time (c. 475 Ma) that the first signs of life appeared on land in the form of very simple microscopic plants. Steur comments:

> The oldest indications for the existence of real land plants have been found from boreholes in Oman. They contained fours of mutually connected spores (tetrads) enveloped by remains of the spore sac in which they had been formed. Research on the spore walls point to a relationship with the liverworts. The fossils have been found in the Middle Ordovician and are about 475 million years old.[85]

It was about fifty million years later (c. 425 Ma) during the middle Silurian Period that *macroscopic* land plants first appeared as evidenced by the fossil record. Steur asserts:

> The oldest fossils of land plants visible with the naked eye are about 425 million years old, from the Middle Silurian. From this time on the plants spread over the land and the continents turned to green. This was the beginning of an amazing development, which created the terms for animal life on land.[86]

The fossils of these Silurian plants (e.g., Cooksonia) are extant spores found within larger plant fragments showing an increasing complexity of the terrestrial plant kingdom over a relatively short period of time.[87] It was not, however, until the late Devonian Period that the terrestrial plant kingdom began to take on a whole new shape. It was during this time (c. 370 Ma) that the first modern tree—the Archaeopteris—appeared. It was found heavily in Euramerica, Siberia, China, and Gondwana. Archaeopteris became the dominant member of the Devonian flora and "the lynchpin of the first true forests."[88] It was also during the late Devonian that the first seed plants appeared (c. 385 Ma) and began to flourish. The fossilized remains are

---

85. Steur, "How Plants Conquered the Land," 1. For further, see also Heckman et al., "Molecular evidence," 1129–33.
86. Steur, "Fossil Plants."
87. Wellman et al., "Fragments of the earliest land plants," 282–85.
88. Murphy, "Archaeopteris," 1.

typically ovules or cupule fragments. It is thought that these fragments are what is left of small shrubs and vines.[89]

The Suggested Solution

We posit that Genesis 1:9–13 can plausibly be understood to refer, not literally to the very first rocks or very first land that raised its head above the primordial seas, but rather to the first land that was permanently inhabitable and conducive to the appearance of terrestrial life. In verse nine, note that God did not say, "Let the land appear," but rather he said, "Let the *dry* land (Hebrew *yabbashah*, lit., "the dry") appear." It is a very specific and important qualifying remark. This implies some sort of new permanence coming from a prior situation of constant change. All land that had previously appeared before the middle Ordovician was not typical dry land (as we know it), but a contrasting form of unstable land; that is, barren and desolate land enveloped with volcanism and earthquakes and constantly sinking back under sea-level or into the molten realm and under continuous and rapid modification. Remember White's description of the prior tectonic conditions:

> [E]verything happened a good deal faster. The magma on which the continents floated was hotter, less viscous, and closer to the surface. Hot spots were probably hotter. The continents moved swiftly, collided more often and tended to fracture or suture with greater frequency.[90]

In other words, up to Day Three, the land situation was very different; it had not yet been given its current equilibrium. Before Day Three, any land that existed was inundated with conditions that could not yet support true land life. The Mosaic description of "dry" land means that God had then taken a new action and crafted the land into an environment of stability and potential abundance. Note that "the dry" (Hebrew *yabbashah*[91]) is a very common Hebraism referring to *solid* ground,[92] i.e., a picture of constant stability. This conclusion is further supported by the fact that these

---

89. Murphy, "Early Seed Plants," 1, 3.
90. White, "Proterozoic," 1.
91. Brown et al., *Hebrew and English Lexicon*, "yabbashah," 387 (H3004).
92. Schrader, 'Genesis," 11. Note that the word "land" is not in the original text; it is, however, implied by *yabbashah*.

two key actions by God on Day Three, i.e., the making of solid and stable dry land with abundant and profusely increasing land plant life, are very intentionally connected and presented together. The new land included both a new stability and a new inhabitability; the stability of the land, in fact, provided for the inhabitability of the land. The Scriptures clearly show that God was engaging in very purposeful and life-centered land development. This postulation fits quite well within the context of the biblical Anthropic Principle.

GOD DAY FOUR

Introduction

In our model, we have included the Carboniferous Period (359.2–299 Ma), the Permian Period (299–251 Ma), and the Triassic Period (251–199.6 Ma) as God Day Four. The Carboniferous Period and the Permian Period was a time of explosion for terrestrial plant life. Specifically, the Carboniferous Period was characterized by the massive dominance of fern forests, while the Permian Period was characterized by the massive dominance of gymnosperm forests. In fact, so great was the proliferation of the world flora during this time that the Earth's atmosphere was significantly changed due to the incredibly increased oxygenation. It should also be noted, from a pure geological perspective, that the Carboniferous (lit., "coal-bearing") is marked worldwide by definitive and rich continental carbon deposits, yet having a particular clarity in the North American stratigraphy with the Mississippian (limestone/carbonate beds) and Pennsylvanian (coal/carbon beds) subsystems. Moreover, there was also much orogenic activity during this time as tectonic collisions occurred during the formation of the Earth's last supercontinent, Pangaea (completed c. 300 Ma).

Included also in this Day is the Triassic Period, which followed the Carboniferous-Permian, and is marked initially by the Permo-Triassic (P-T) Extinction and then the first appearance of the dinosaurs (251 Ma). Additionally, modern science asserts that proto-mammals first appeared during the late Triassic (c. 220 Ma), having developed from synapsids (which first appear about 310 Ma: the earliest and more reptile-like synapsids are sometimes called pelycosaurs; the more recent and more mammal-like are called therapsids).[93] In this paradigm, modern mammals are

---

93. Stearn and Carroll, *Paleontology*, 265–75. Note the sections, "Mammal-Like

considered to be the evolutionary descendants of the therapsidic lineage via mammalian cynodonts.[94] However, if understood within the context of the scriptural revelation and a non-evolutionary interpretation, these variant Paleozoic/Mesozoic creatures cannot be construed to be either true mammals or morphologically ancestral to mammals. Regardless of any biological similarities, true *nephesh* mammals did not appear until late God Day Five and early God Day Six, which in our model falls into alignment with the Cenozoic Era.

Reptiles" (265–70) and "Mesozoic Mammals" (270–75). They hold that mammalian ancestry goes back to the late Paleozoic.

94. For representative sampling of proposed evolutionary mammalian sequencing, see Cifelli, "Early Mammalian Radiation," 1214–26.

The Correlation Model

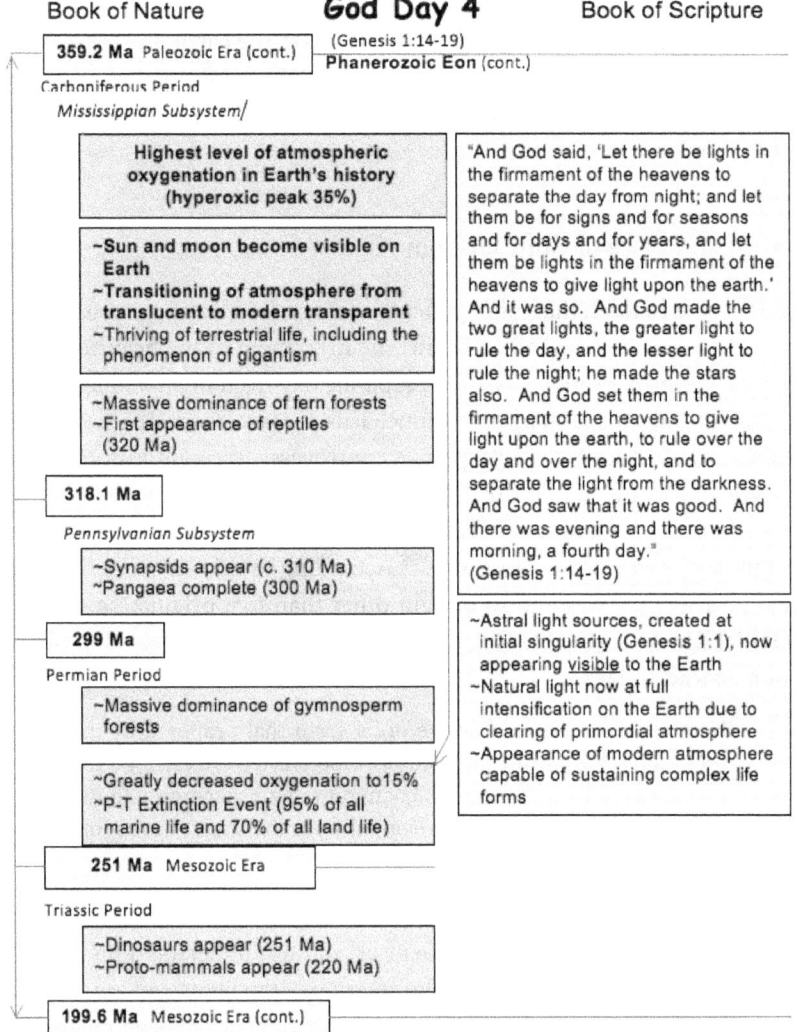

Figure 7: God Day Four correlation

The Scriptural Text: God Day Four

Genesis 1:14–19 states: "And God said, 'Let there be lights in the firmament of the heavens to separate the day from the night; and let them be for signs and for seasons and for days and for years, and let them be lights in the firmament of the heavens to give light upon the earth.' And it was so. And God made the two great lights, the greater light to rule the day, and the lesser light to rule the night; he made the stars also. And God set them in the firmament of the heavens to give light upon the earth, to rule over the day and over the night, and to separate the light from the darkness. And God saw that it was good. And there was evening and there was morning a fourth day."

The Significance of God Day Four

The Scriptures teach that the significant event of God Day Four was the first appearance of the heavenly lights in "the firmament of the heavens" in order *"to give light* upon the earth," to separate day from night and light from darkness, and for the marking of time. This passage, Genesis 1:14–19, has been a constant source of interpretive controversy in recent history within the circle of evangelical scholarship. On the one hand, there are those like Schrader, who believe that "These verses record the formation of the sun, moon, and stars on the fourth day."[95] According to this view, the light created on Day One was a form of light other than that originating with the celestial lights because the sun, moon, and stars were not created until Day Four. Schrader claims:

> Here again is a reversal of the order from that proposed by the evolutionists. According to Genesis, God created the earth on the first day, and then the sun on day number four. According to the evolutionist, the earth was thrown off from the sun or bore some other relationship to the sun.[96]

On the other hand, there are those like Sailhamer, who believe that "it [the text] suggests that the author [of Genesis] did not understand his account of the fourth day as an account of the creation of the lights; but, on the contrary, the narrative assumes that the heavenly lights have been created already 'in the beginning.'"[97] According to this view, the lights were recorded as having been created in Genesis 1:1 (the creation of the heavens and the earth), yet only came into Earth's clear view in Genesis 1:14–19 (Day Four).

From a textual perspective, the key to determining which view is best is centered around the meaning of the phrase "the heavens and the earth" (Hebrew *shamayim eth erets*) in Genesis 1:1. In agreement with Sailhamer,

---

95. Schrader, "Genesis," 12.
96. Ibid.
97. Sailhamer, "Genesis," 34.

we have already asserted our understanding of Genesis 1:1 as referring to the creation of the entire general cosmos, which includes everything that exists outside of God and the supernatural realm. This, of course, means that the sun, moon, and stars existed long before Creation Day Four. Interestingly, Sailhamer also suggests that, in the Hebrew, the syntax of verse six (Day Two) affirms the divine crafting of the firmament whereas the syntax of verse fourteen (Day Four) affirms the divine *separation* of the lights in the firmament (as opposed to the *creation* of the lights in the firmament).[98] If the Sailhamer interpretation is correct (and we believe that it is), it implies that the lights were already there. Nonetheless, the most important aspect is certainly the *ex nihilo* creation of the lights along with the rest of the material universe in Genesis 1:1—*prior* to the Creation Week.

This view also fits much more cohesively with the postulations of modern science. Whereas Schrader's view finds itself in direct conflict with cosmology (of which he erroneously lumps and equates with evolutionism), our view does not. In our model, we hold that the significant event as recorded in Scripture (the appearance of the sun, moon, and stars) is a direct result of the final clearing of the Earth's atmosphere from translucent to transparent. This event occurred in the Carboniferous Period and into the Permian Period due primarily to the immense increase of plant-related oxygenation, particularly from that of the giant forests that dominated the land.

According to Kazlev: "So vigorous is the growth of these ancient trees that they seemed to have sucked much of the carbon dioxide out of the atmosphere, producing a surfeit of oxygen. Oxygen levels were higher during this time than any other time in the history of the Earth."[99] Pavle I. Premovic adds "that atmospheric O_2 increased from approximately 18% to 21% between the middle and late Devonian (ca. 380–360 Ma) and then rose sharply to about 35% by the late Carboniferous-early Permian (ca. 290 Ma), a remarkable value compared with the present atmospheric level (PAL) of 21%."[100] Science posits that the appearance and subsequent explosion of vascular plants during this period of time with their increased production of oxygen, coupled with the resultant mass burial of organic carbon as the plants died, was the catalyst. The atmospheric oxygenation was greatly

---

98. Ibid.
99. Kazlev, "Carboniferous," 4.
100. Premovic, "Late Paleozoic oxygen pulse," 143.

intensified as the increased burial of organic carbon further accelerated the growth of very large woody vascular plants and other expansive foliage.

The Carboniferous landscape was heavily inundated with swamps. In time, as the giant trees and ferns died, they sunk into the swamps and were quickly buried before bacteria could bring about their decomposition (the process of decomposition absorbs oxygen). This continual process formed extensive peat beds which eventually turned into what is observed today as Carboniferous coal. The result was a near doubling of the atmospheric oxygen levels from that of the pre-Carboniferous. As Robert A. Berner explains:

> The plants supplied a new source of organic matter to be buried on land and carried to the oceans via rivers. This "new" carbon was added to that already being buried in the oceans, thus increasing the global burial flux. This is especially true of lignin, a substance that is decomposed only with difficulty by micro-organisms. The rise of ligniferous plants and an initial level of microbial lignin breakdown lower than that at the present may have contributed to increased organic matter burial and better preservation. This high burial rate is reflected by the abundance of coals during this period, which is the greatest abundance in all of earth history.[101]

It is thought that the 35 percent oxygen production is the highest level possible without being devastating to the plants themselves (since plants breathe carbon dioxide). The Carboniferous would have been a time of great atmospheric balancing. As Berner advises:

> One problem with high levels of atmospheric oxygen is that this situation should be deleterious to plant life, especially if atmospheric CO2 is not elevated as is believed to be the situation for the late Carboniferous . . . Nevertheless . . . [it has been] shown that a process-based terrestrial carbon cycle model, based on plant growth experiments, predicts that an atmospheric composition of 35% O2 and 300 ppm [parts per million] CO2 does not invalidate terrestrial ecosystem biogeochemical cycling of carbon, but simply places limits on photosynthetic productivity and canopy structure at the global scale.[102]

One must not miss the significance of this concept. From an evangelical Christian worldview, this can be appropriately interpreted as evidence of

---

101. Berner, "Atmospheric oxygen over Phanerozoic time," 10956.
102. Ibid., 10957.

The Correlation Model

the providential hand of God at work in the detailed conservation of his bioproductive environment. God has provided a system of checks and balances that enable his creation to function within definitive life-conducive perimeters.

The Carboniferous and Permian periods were a time of great thriving for all terrestrial life. With the massive amplification of vascular plant growth along with the resultant 35 percent peak oxygenation of the atmosphere, it should be noted that animal life benefited as well. Scientific studies have suggested that the oxygen-flooded atmosphere resulted in the gigantism of many species of animals, namely insects, amphibians, and other arthropods. Researchers have long since marveled at the size of many insects found in the fossil record at certain periods of Earth's geologic history (e.g., giant dragonflies, like Meganeura), particularly in the Carboniferous and early Permian strata. It seems very probable that the coinciding of the animal gigantism with the times of high atmospheric oxygenation is not accidental.[103]

By the late Permian Period, the oxygen level had greatly dropped from the 35 percent hyperoxic peak during the late Carboniferous to about 15 percent. This change (a new hypoxia) is often thought to have been at least one of several major factors in the End-Permian (P-T) Extinction Event—the largest extinction in Earth's history (of which 95 percent of all marine life and 70 percent of all terrestrial life perished).[104] According to Robert Dudley:

> The severe end-Permian extinctions of both terrestrial and marine taxa have been, in part, attributed to anoxic conditions, although a diversity of biotic and abiotic factors may have contributed synergistically to this effect. The disappearance of giant terrestrial arthropods with diffusion-limited respiratory systems is, however, consistent with the causal mechanism of atmospheric hypoxia restricting such taxa to progressively smaller body sizes. Similar conclusions apply to the giant semiaquatic and terrestrial amphibians of the late Paleozoic that became extinct by the end of the Permian.[105]

103. Monastersky, "Ancient animals got a rise out of oxygen," 1–2.

104. Goodwin et al., "The Permo-Triassic Mass Extinction." Goodwin lists several possibilities including volcanism, impact, and climate change. It is probable that hypoxia could have resulted from a combination of these causes. There is strong evidence of volcanism (e.g., the Siberian Traps) and some evidence of possible impact (e.g., there is both the presence of iridium spikes indicative of impact radiation and the presence of fungal cells indicative of catastrophic vegetation breakdown) at the P-T boundary.

105. Dudley, "Atmospheric Oxygen, Giant Paleozoic Insects," 1045.

Despite the severe oxygen crash, it is important to note, however, that the massive Carboniferous oxygenation had permanently affected the Earth's veil clearing it of residual primordial remnants and serving as the final cog of the modern atmosphere. There would be another milder, but significant, rise of oxygenation (to about 25–27 percent) in the mid-Jurassic Period that would last through the Cretaceous and for most of the Tertiary Period (God Day Five). Eventually, the oxygenation would level out in the Tertiary Period to the present day normoxic 20.9 percent.[106] This current oxygen level is ideal for the existence of today's modern life.

The natural record shows that the dinosaurs first appeared at the Permian-Triassic boundary. This is indeed considered to be a significant event to modern science, but not to the Holy Scriptures. It is important to note that this does not indicate a contradiction between the scriptural revelation and the natural revelation. It serves as a reminder that the Genesis creation text is not exhaustive in detail, but rather shows a progression of high points toward God's ultimate purpose. It also serves as a reminder that there are some things that rightly grab the extreme investigative interest and curiosity of humanity (such as dinosaurs). Yet, some of these things (such as dinosaurs), while certainly having their particular place in the progression of the natural order, may not have the same place or degree of importance in the greater divine plan as do other things.[107] Sometimes the fascinations of humanity and the proclamations of God are substantively different.

GOD DAY FIVE

Introduction

In our model, God Day Five consists of the Jurassic Period (199.6–145.5 Ma), the Cretaceous Period (145.5–65.5 Ma), and part of the Paleogene

---

106. Ibid., 1044. Note: Ross also includes the slowing of the Earth's rotation rate and the carbonate modification factor as influences in the changing of the atmosphere (See Ross, *The Genesis Question*, 41–43).

107. It is interesting that dinosaurs have been given such a great place within the worldview of naturalistic evolutionism, but are not overtly mentioned (if indeed mentioned at all) in the scriptural revelation. This could and should be understood as a significant message to those who try to incorporate macroevolutionary theory into a biblical schematic. It should also be noted, however, that the dinosaurs have provided the primary foundation for the existence of fossil fuels. This supports the metanarrative that is the divine Anthropic Principle.

The Correlation Model

Period (65.5–53.5 Ma; specifically, the Paleocene Epoch and a small portion [2.3 myr] of the early Eocene Epoch). According to science, the dinosaurs thrive and dominate the Earth for the major portion of this Day—until the K-T boundary (65.5 Ma). However, from the scriptural perspective, the key events on this Day do not include the dinosaurs, but rather the creation of swarms of sea life (esp., aquatic mammals) and the advent of aerial life (esp., true birds[108]).

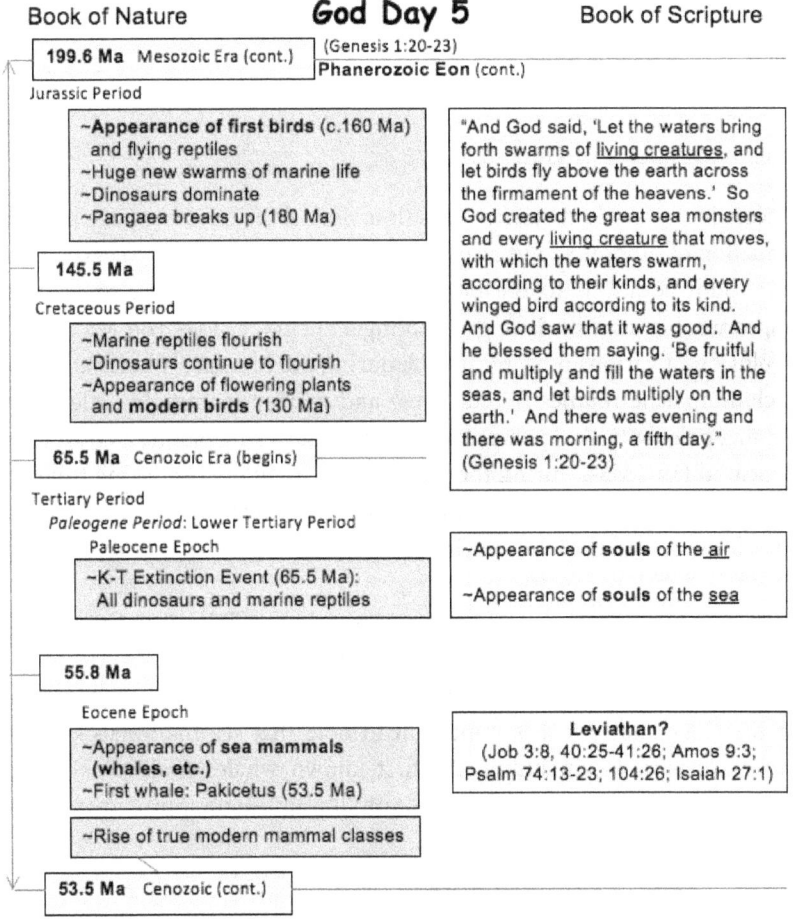

Figure 8: God Day Five correlation

108. Xu et al., "An *Archaeopteryx*-like theropod," 465–70. It seems that the recent discovery of Xiaotingia zhengi (c. 160) in Liaoning, China may now displace Archaeopteryx (c. 150 Ma) as the first known true bird. Also, keep in mind that there is much ongoing discussion within the circles of paleobiology as to what defines a *bird*. For instance, see Tamura, "Xiaotingia zhengi," 1–3.

The Scriptural Text: God Day Five

Genesis 1:20–23 states: "And God said, 'Let the waters bring forth swarms of living creatures, and let birds fly above the earth across the firmament of the heavens.' So God created the great sea monsters and every living creature that moves, with which the waters swarm, according to their kinds, and every winged bird according to its kind. And God saw that it was good. And he blessed them saying, 'Be fruitful and multiply and fill the waters in the seas, and let birds multiply on the earth.' And there was evening and there was morning, a fifth day."

The Significance of God Day Five

According to modern science, this time period includes not only the continuation of dinosaurs, but also their absolute domination. However, other important additions to the biosphere occurred as well: the first known appearance of true birds (e.g., Xiaotingia zhengi, c. 155–160 Ma) and an explosion of various new marine life during the Jurassic.[109] Later additions include marine reptiles (Cretaceous) and marine mammals (Paleogene). From a paleobotanical perspective, the first flowering plants, angiosperms, appeared (130 Ma—during the Cretaceous). At about that same time, the first modern birds (Aves) appear as well. Also very significant, from the mid-Jurassic and throughout the Tertiary, there was another strong pulse of atmospheric oxygenation.

Dinosaurs dominated the entire Mesozoic Era (Triassic to Cretaceous). After the extinction of the dinosaurs and marine reptiles (65.5 Ma—at the K-T Extinction Event), mammals ultimately became the dominant animals in the Cenozoic Era. It is important to note that sea mammals—that is, archaic cetaceans—including the first known whales (Pakicetus), based on current paleontological findings, initially appeared approximately 53.5

---

109. We could place the Day Four-Day Five boundary specifically at 155–160 Ma, which is the appearance in the stratigraphic record of Xiaotingia zhengi (earliest known bird). Comparatively, science places Crown Aves (modern birds) at 130 Ma. However, it is quite probable that the radiation of Avialae goes deeper in time than 160 Ma., so—since there is no remarkable event precluding this for our model—and in order to give some ample room for earlier birds to be found, we used the official Jurassic boundary as the beginning point of this God Day. Lee suggests the possibility that Paravians (basal birds) could extend back as early as 200 Ma., which is just below the base boundary of the lower Jurassic (199.6 Ma). Perhaps time will tell. See Lee et al., "Morphological Clocks," 443.

million years ago during the Eocene Epoch (Paleogene Period).[110] It is reasonable to believe that certain ancient cetaceans (e.g., whales) may be the "Leviathan" of the Old Testament.[111] Many of these creatures, though smaller in stature than modern day whales, were much more fierce and "dragon-like" in appearance. Moreover, the paleontological record shows that many early cetaceans had fully developed legs and were both terrestrial and aquatic.[112] In addition to Pakicetus (53.5 Ma), these early whales prevailed across the majority duration of Eocene time. Note the chronological sequence: Ambulocetus (50 Ma), Maiacetus (48 Ma), Rodhocetus (47 Ma), Indocetus (45 Ma), Protocetus (42 Ma), Dorudon (40 Ma), Basilosaurus (35 Ma), etc.[113]

Day Five includes the significant Mesozoic-Cenozoic transition. The Mesozoic Era is known as "the age of reptiles," while the Cenozoic Era is known as "the age of mammals." Dinosaurs and their ilk appear, dominate, and then perish from the Earth. Meanwhile, as the reptilian dominance faded with the end of the dinosaurs, true modern mammals appear and increase (relegating extant reptiles to lesser status), and then persevere in strong dominant fashion to the present time.

According to the Scriptures, God Day Five is the first of two Days in which God created *living souls* (Hebrew *chay nephesh* = living souls, living persons, living selves: *chay* = alive, living[114]; *nephesh* = soul, being, self, person[115]). On Day Five, God created the *living souls* (Hebrew *chay nephesh*) of the sea (Hebrew *sherets* = aquatic swarming things[116]) and the air (Hebrew

---

110. Bajpai and Gingerich, "A new Eocene archaeocete," 15464–68. Specifically, this genus of the Pakicetidae family is called Himalayacetus.

111. Some have tried to connect "Leviathan" (Hebrew *livyatan*) to the dinosaurs, but such a connection is uncertain and we believe improbable. Leviathan is mentioned in five places in the Bible (Job 3:8; 41:1–34; Pss 74:14; 104:24–26; Isa 27:1) and is most often associated with "sea monsters," or more probable, ancient whales.

112. Many scientists interpret this as an evidence of biological macroevolution. See Sutera, "The Origin of Whales and the Power of Independent Evidence," 4. See also, Thewissen and Bajpai, "Whale Origins,"1037–49.

113. Note the frequent use of the Greek *ketos* (i.e., English transliteration "cetus") = "whale," or "huge sea monster."

114. Brown et al., *Hebrew and English Lexicon*, "chay," 311 (H2416).

115. Ibid., "nephesh," 659 (H5315).

116. Ibid., "sherets," 1056 (H8318).

owph umph = flying creatures, fowl: *owph* = fowl[117]; *umph* = fly, hover[118]). (On Day Six, God created the *living souls* of the land—see next section.)

This emphasizes an important qualitative distinction; there are greater and lesser creatures. Not all creatures are nephesh. There are those creatures specifically designated by God as being *living souls* and there are other creatures that are not. The failure to mention the dinosaurs in this capacity (indeed, possibly failing to mention them at all in the Scriptures[119]) is indicative that they were not created as *living souls*. This does not mean that such creatures did not "live" in a biological sense. The fossilized remains of dinosaurs are certainly evidence of their prior existence as biological organic beings. It simply means that only certain creatures were given a *soulish* dimension by God; others (notably dinosaurs, insects, protozoans, etc.) were not.[120] Note that plants, while considered to be a biological life form, have not been gifted with the same inherent life *value* as have animals, particularly *nephesh* animals.

Another very significant matter that reinforces the high place of nephesh creatures is that of divine blessing. On Day Five, God pronounced his blessing specifically on the nephesh creatures. Sailhamer explains:

> For the first time in the creation account, the notion of "blessing" appears (v. 22). The blessing of the creatures of the sea and sky is identical with the blessing of man, with the exception of the notion of "dominion," which is given only to man. As soon as "living beings" [*chay nephesh*] are created, the notion of "blessing" is appropriate because the blessing relates to the giving of [soulish] life.[121]

---

117. Ibid., "owph," 733 (H5775).

118. Ibid., "umph," 733 (H5774).

119. It has also been claimed that the *behemoth* (Hebrew *behemowth*; H930), described in Job 40:15–24, refers to a dinosaur, possibly a brachiosaur or a diplodocus. However, there is no way to make any specific determination. Nonetheless, the text is definitively a description by God himself of some sort of primitive creature (v. 19a—"He is the first of the works of God" = Hebraism for "ancient" [even prehistoric]), but *without any reference whatsoever to an inherent soulishness*. This is plausibly indicative that the behemoth (whatever it may have been) was not a nephesh creature. (NOTE: Higher critical scholars often interpret both behemoth and leviathan, not as historical creatures at all, but as mythical amplifications of evil and chaos.)

120. For a contrasting view, see Leupold, *Exposition of Genesis I*, 83. He holds that nephesh is merely an "animating principle" possessed by all animals.

121. Sailhamer, "Genesis," 35.

The Correlation Model

God gives his special blessing to all creatures that have received from him the gift of nephesh life. Soulish beings are of a higher order of existence than all others. Though science shows that lower life forms have existed since the Archean Eon (c. 3.5 Ga), which we place in God Day One, nephesh creatures did not appear until much more recently in God Day Five. Please note that, with such reasoning in mind, we do not believe that the scriptural text of Genesis 1:20–22 is intended to be inclusive of all sea creatures, nor all sky creatures. The text is narrowly descriptive of "great sea monsters" (Hebrew *gadowl tanniyn*: *gadowl* = great [in magnitude and extent, in number, in intensity and fear, in sound][122]; *tanniyn* = sea monster, serpent, dragon, personification of water spout[123]) that move in the waters and "winged birds" (Hebrew *kanaph owph*: *kanaph* = wing, extremity[124]) that fly in the sky. Contextually, this does not include, for instance, fish, plankton, and flying insects—namely, lower life forms. We believe that the text refers to higher beings like marine mammals and birds—namely, creatures that are soulish. Such mammals and birds have the unique capacity to engage in complex social and emotional relationships, often even with human beings. Therefore, on Day Five, God is not merely filling the Earth with more creatures; indeed, he has been doing that from very early on (e.g., first fish appeared c. 535 Ma as a late part of the Cambrian profusion: i.e., long before Day Five); the point of the text is that God is now filling the Earth with nephesh beings. He is moving progressively toward his ultimate creation, humanity, who appears on Day Six. With this understanding, one sees the clear commensuration of the biblical ordering of events with the sequence and time-line of natural revelation.

---

122. Brown et al., *Hebrew and English Lexicon*, "gadowl," 152 (H1419).

123. Ibid., "tanniyn," 1072 (H8577). In addition to this text (Genesis 1:21 = sea monsters), the Bible uses the term in various other ways (Exod 7:9–12 = serpent; Isa 51:9 and Ezek 29:3 = dragon). Hebrew *tanniyn* is a term that is generally phenomenological and not truly taxonomic; this leaves it open to contextual interpretation. We believe that its usage typically signifies creatures that are not mundane or ordinary, but rather having some form of great significance, power, or fearsomeness that is determinate by the context.

124. Ibid., "kanaph," 489 (H3671).

GOD DAY SIX

Introduction

God Day Six is the culmination of the divine creative sequence. On this Day, God created first the nephesh of the land and then, finally, humanity. In our model, parts of the Tertiary Period—specifically, the latest (and vast majority) portion of the Eocene Epoch (53.5–33.9 Ma), the Oligocene Epoch (33.9–23.0 Ma), the Miocene Epoch (23.0–5.33 Ma), and the Pliocene Epoch (5.33–2.588 Ma)—as well as part of the Quaternary Period, namely, from the beginning of the Pleistocene Epoch (2.588 Ma) up to the 1.9 million year point, are considered to be God Day Six. There are two important points to mention.

First, the Eocene Epoch is the rough transitional period between Day Five and Day Six in our correlation model. We have divided it out of necessity at roughly 53.5 Ma because that is where science currently confirms the appearance of Pakicetus,[125] the first known whale taxa (thus, concluding Day Five in our model with the earliest known appearance of aquatic mammals). However, there is some overlap because, according to the paleontological record, modern land mammals also existed throughout the entire Eocene (which begins at 55.8 Ma), i.e., just 2.3 million years prior to Pakicetus, while the scriptural record seems to present marine nephesh appearing prior to land nephesh. Though this is no major issue, it does make a precise Day Five-Day Six boundary delineation slightly uncertain among *known* empirical markers. Of course, this difficulty would completely clear up if cetacean fossils were ever to be discovered either at the Eocene boundary, or earlier in Paleocene strata (which we believe will happen). As such, we remain vigilant in awaiting for that eventuality to occur.[126] Indeed, we

---

125. Bajpai and Gingerich, "New Eocene Archaeocete," 15464–68. (Specifically, this Pakicetid is called Himalayacetus.)

126. Ibid. Bajpai makes this *projective* assertion: "When the temporal range of Archaeoceti [early whales] is calibrated radiometrically, comparison of likelihoods constrains the time of origin of Archaeoceti and hence Cetacea to about 54–55 Ma (beginning of the Eocene), whereas their divergence from extant Artiodactyla may have been as early as 64–65 Ma (beginning of the Cenozoic)." He is projecting that whales most likely existed near the beginning of the Eocene Epoch and probably even much earlier, perhaps first appearing as early as the lower Tertiary Period. In fact, using projective calculations, Bajpai favors placing Pakicetus at 56.7 Ma, which, if correct (and we believe that he is), would ideally fit our correlation timeline and immediately create an empirically precise Day Five-Day Six boundary.

The Correlation Model

are certain that the Archaeoceti are there, in upper Paleocene rock, just waiting to be found.

Second, Day Six is concluded at 1.9 Ma because it is our assertion that true humanity first appeared in the record at about that time. This is the approximate point on the time-line that modern anthropology places the appearance of *Homo erectus*. We strongly suggest that *Homo erectus* is actually the beginning of early modern *Homo sapiens* (i.e., Adam). The appearance of *Homo erectus* is also just prior to the Gelasian-Calabrian boundary (1.8 Ma) of the Pleistocene Epoch (which, until June 2009, was considered to be the Pliocene-Pleistocene epochal boundary[127]); this is also very significant. Most Christian scholars, even Old-Earth creationists, refuse to go back that far in time to allow for the first appearance of true humanity. In fact, most scholars will only go back to (at most, but usually less than) 200,000 years for *Homo sapiens sapiens* (which is the classification of all people today).[128] Ross, for instance, places original humanity back to only approximately 37 to 50 thousand years ago.[129] However, in our model, the 1.9 million year ago placement is not arbitrary and has significant connection to other major events of history, including the Noahic Flood and the great Pleistocene glaciation (the traditional Ice Age). From an evangelical Christian perspective, and in consideration of an integrationist approach to truth, all of these events must be taken seriously.

---

127. Gibbard et al., "Formal Ratification of the Quaternary System," 96–102.
128. O'Neil, "Early Modern Homo Sapiens," 1.
129. Ross, *Matter of Days*, 224–25.

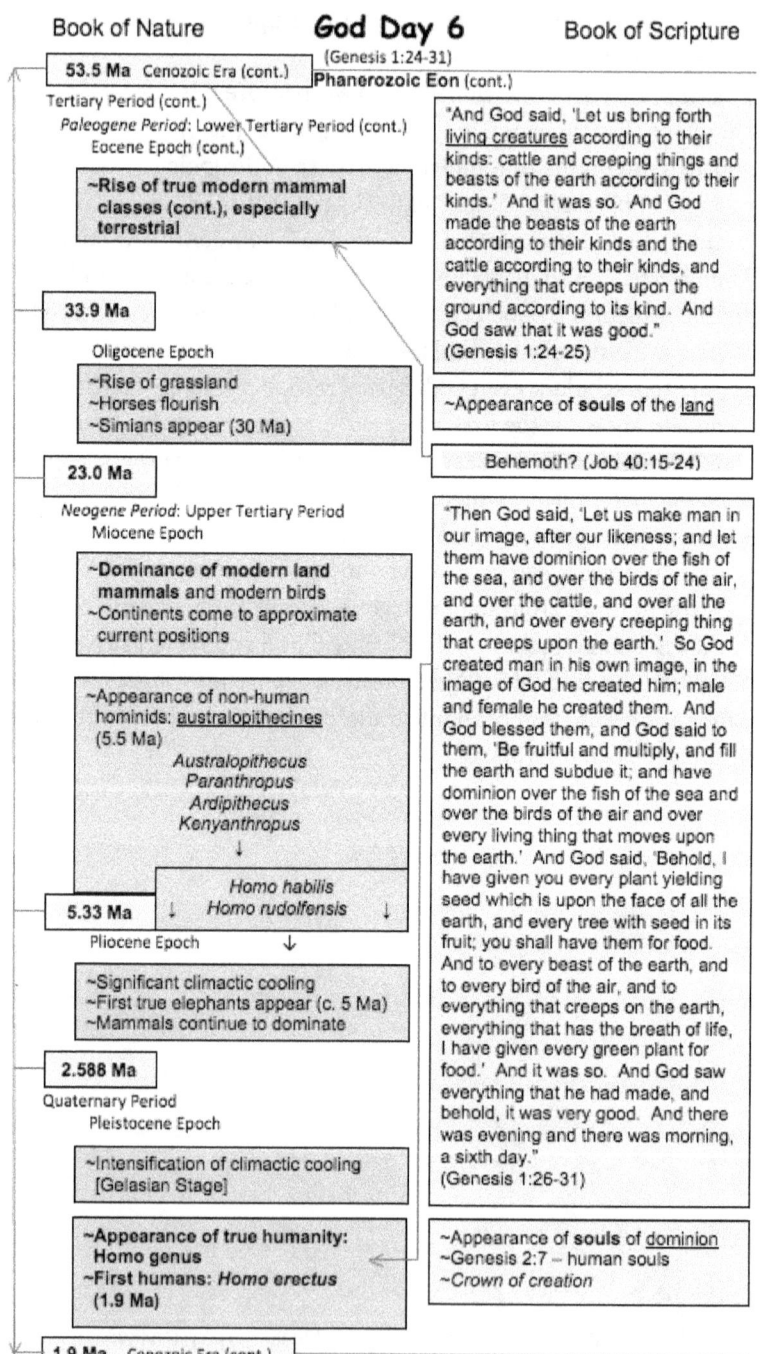

Figure 9: God Day Six correlation

The Correlation Model

The Scriptural Text: God Day Six

Genesis 1:24–25 states: "And God said, 'Let the earth bring forth living creatures according to their kinds: cattle and creeping things and beasts of the earth according to their kinds.' And it was so. And God made the beasts of the earth according to their kinds and the cattle according to their kinds, and everything that creeps upon the ground according to its kind. And God saw that it was good."

Genesis 1:26–31 continues: "Then God said, 'Let us make man in our image, after our likeness; and let them have dominion over the fish of the sea, and over the birds of the air, and over the cattle, and over all the earth, and over every creeping thing that creeps upon the earth.' So God created man in his own image, in the image of God he created him; male and female he created them. And God blessed them, and God said to them, 'Be fruitful and multiply, and fill the earth and subdue it; and have dominion over the fish of the sea and over the birds of the air and over every living thing that moves upon the earth.' And God said, 'Behold, I have given you every plant yielding seed which is upon the face of all the earth, and every tree with seed in its fruit; you shall have them for food. And to every beast of the earth, and to every bird of the air, and to everything that creeps on the earth, everything that has the breath of life, I have given every green plant for food.' And it was so. And God saw everything that he had made, and behold, it was very good. And there was evening and there was morning, a sixth day."

The Significance of God Day Six

The key to understanding God Day Six is three-fold. First, God completed the establishment of soulish dominance of the Earth by filling the *land* with soulish creatures. (He had previously established soulish dominance of the *sea* and *sky* on Day Five). This signifies that the entire Earth—sea, air, and land—had been filled with the greater creatures of God.

Second, God crowned his creation with humanity, the soulish creature bearing his own image. The creation of humanity (1:26–27) served as the personal imprint of divine authorship and care. The blessing of humanity (1:28) served as the personal consecration and seal of divine approval.

Third, God established the premise for the benevolent care of the created order (1:29–30). This is connected to the human dominion mandate with further qualitative distinctions made. Life was created to reign supreme over non-life; animal life, particularly nephesh life, was created to reign supreme over plant life. The non-living physical order and the plant kingdom, (i.e., non-sentient life with no life blood—see Gen 9:3–6) was created to be a provision for the existence of the sentient-living physical order. Moreover, the creatures of the living physical order were created to peacefully co-exist with one another in relational companionship and not in a "red in tooth and claw" struggle for survival. After all (contrary to the majority view of OEC, including Ross), in the original pre-fall creation, plants (not animals) were fiat given as the divine provision for human and animal sustenance (Gen 1:30). Finally, humanity was given the privilege and responsibility to be the benevolent caretakers (not the destructive abusers) of the physical order.

The scripture text (Gen 1:24–25) reveals that God created the nephesh of the land on Day Six. There are three terms used in the passage—two are more specific and one is more general. In orderly succession, the terms are as follows: "cattle" (Hebrew *behemah*), "creeping things" (Hebrew *remes*), and "beasts of the earth" (Hebrew *chay erets*). In actuality, the word *behemah* typically refers to larger animals, usually those domesticated such as cattle, horses, or mules—but sometimes referring to wild carnivores[130]; the word *remes* typically refers to "small animals"[131]; finally, *chay erets* is a general term that simply refers to "land animals." This usage of words is significant. While the terms themselves are not taxonomically exhaustive or extremely specific, they are not a mere random sampling of animal types. We believe that the author is conveying a definitive contextual picture. Since each of these creatures are nephesh beings (Gen 1:24), we suggest that the grouping systemically refers to *all* land mammals—wild and domesticated, large and small.[132] With this presentation of nephesh *land* appearance, the scriptures communicate the final divine establishment of nephesh animal dominance of the Earth (having already established this dominance of sea and air on God Day Five).

---

130. Brown et al., *Hebrew and English Lexicon*, "behemah," 96 (H929).
131. Ibid., "remes," 943 (H7431).
132. For a good discussion of this, see Schicatano, *Theory of Creation*, 93–101.

THE CORRELATION MODEL

Finally, God crowned the universe with the creation of humanity (the ultimate nephesh = bearing divine likeness); this was God's grand statement of completion. He put his divine image in people. Humanity was given both a natural (i.e., emotions, will, and intellect) and moral (i.e., ability to know and discern right and wrong) likeness to the Creator. This differentiation sets human beings qualitatively apart from and higher than all other nephesh beings, including ape primates and non-human hominids.[133] While nephesh beings are on a qualitatively higher plain (i.e., possessors of living souls) than non-nephesh creatures, human beings are likewise qualitatively higher than all other nephesh beings (i.e., bearers of the image of God) (cf. Matt 10:29–30; 12:12). God created humanity as the *souls of dominion*.

In contrast, anthropology often struggles with the alleged evolutionary relationships of the various hominid taxa. This is particularly true concerning how they are perceived to relate to *Homo sapiens*. Of course, anthropology does not typically give consideration to the metaphysical dimensions of reality. Thus, the evolutionary assumptions of modern anthropology provide no spiritual differentiation between the various mammalian taxa (or, for that matter, between any higher life forms and lower life forms). Such assumptions can lead easily to the understanding that ape primates and australopithecine hominids are the ancestral relatives of modern humanity. After all, the grounds for taxonomic comparison and the development of systemic evolutionary sequencing are merely matters of anatomical and genetic similarities. Even human mental function, indeed life itself, is absurdly considered to be nothing more than chemical processes.[134] This view, however, conflicts with the teaching of Scripture. It disregards the divine image inherently present in humanity and the dominion mandate given by God. There are Christian scholars, such as Stearley, who seem to leave open the possibility that God placed his image on Adam sometime *after* he had been created and had later completed a certain level of evolutionary development (particularly as related to human cognition). He states:

---

133. Stearley, "Assessing Evidences," 169. He states, "Classically, Christian theology holds that the image of God consists of various capacities, such as rationality or a moral sense, which are uniquely possessed by humanity." Note that while certain other creatures do have a degree of rational capacity, it is understood that humanity has been granted the highest level of rationality among living beings.

134. Miller, *Finding Darwin's God*, 168–69. See also Weisz et al., *Science of Biology*, 716–19.

> If the *imago Dei* represents an elective act by the Almighty God to a representational office, based on a cognitive platform designed over time, then may we be permitted to speculate when this appointment occurred? I feel that the common tendency for Christians and non-Christians to focus on a few dramatic benchmarks, such as the eruption of cave art in southern Europe 40,000 BP (the "creative explosion"), misses some more basic, but humble, markers of a significant cognitive platform. For example, Middle Paleolithic culture is typified by the use of fire, multicomponent tools, regional variation in tool production, and human burial. Another potential benchmark might be the origination of anatomically modern humans. A third possibility would be the beginning of Upper Paleolithic culture. Perhaps it is not within our power to discern.[135]

Determining just what point in human development God decided to bestow his image is most certainly "not within our power to discern"—and for very good reason. It is because the whole idea blatantly flies in the face of the biblical teaching. As Schrader rightly comments:

> [F]rom the moment of his creation man was fully capable of exercising control over his environment. It did not take millions of years for man to evolve in his search for self-identity and self-consciousness, for the dominion was the direct consequence of being created in the image of God. Man was created with dominion over all living creatures (v. 26).[136]

The biblical revelation indicates that the dominion mandate was given at the proximate time of Adamic creation. Thus, it is the claim of Scripture that humanity did not come from apes and hominids. Adam—from his initial creation—was given dominion over apes and hominids, and over the entire natural order by divine fiat. With this view, the events of Genesis 2, including the divine creation of a help-mate for Adam and the naming of all the nephesh beings by Adam, makes logical sense. God presented Adam with the other nephesh beings (Gen 2:19; note that the text specifically uses *chay nephesh*) and all were found to be qualitatively deficient for the purpose of the deepest intimate companionship for human beings: "for man there was not found a helper fit for him" (Gen 2:20). This is significant for three reasons. First, it confirms that humanity is of a different and higher

---

135. Stearley, "Assessing Evidences," 169. See also, Finlay, "*Homo divinus*: The ape that bears God's Image."

136. Schrader, "Genesis," 14.

nephesh order than the rest. Second, it should also be understood that only nephesh creatures were presented to Adam in the first place; this is indicative that nephesh beings are higher than non-nephesh creatures. Third, the naming event is an exercise of God-given authority that, by its very occurrence and practice, confirms human dominion over all other nephesh life forms.[137] The advent of humanity signified the divine completion of original creation. This is the essence of God Day Six.

GOD DAY SEVEN

Introduction

In our model, God Day Seven is essentially the time of human existence. It begins at 1.9 million years ago (the approximate point of the advent of humanity), includes the remainder of the Pleistocene Epoch, the entire Holocene Epoch (11.7 ka to present), and proceeds until the Eschaton and the end of time. This Day includes such occurrences as the Edenic Existence, the fall of Adam (Gen 3), the Noahic Flood (which we place 300–1700 years just prior to 1.81 Ma, which is the Gelasian-Calabrian boundary), the Babelian Dispersion, the Ice Age, the Christ Event (2 ka), the present day, and continues until the Eschaton and the end of the age.

---

137. Walton et al., "Genesis," 33.

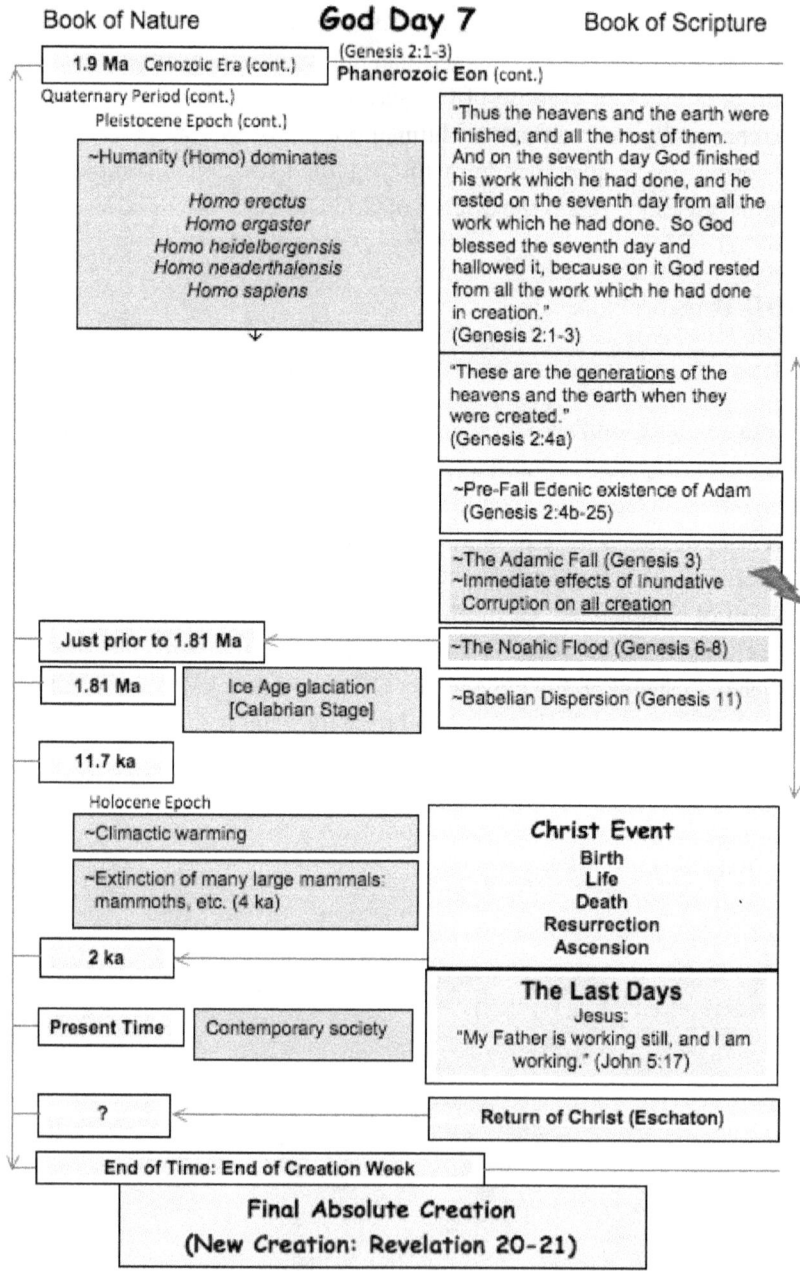

Figure 10: God Day Seven correlation

The Correlation Model

The Scriptural Text: God Day Seven

Genesis 2:1–3 states: "Thus the heavens and the earth were finished, and all the host of them. And on the seventh day God finished his work which he had done, and he rested on the seventh day from all his work which he had done. So God blessed the seventh day and hallowed it, because on it God rested from all his work which he had done in creation."

The Significance of God Day Seven

Since God Day Seven (in many Old-Earth paradigms) is essentially the majority time of human existence, our correlation model holds that it encompasses all chronological time from the proximate point of Adamic creation until the cataclysmic conclusion of the created order via the Eschaton. Evangelical theology, whether from an Old-Earth or Young-Earth perspective, usually places the events of Genesis 2 (2:4b–25) as part of the Sixth Day (i.e., the text is generally understood to be a detailed elaboration of the creation events of Adam and Eve previously recorded in Gen 1:26–27).[138] We agree with this general interpretation—and also that Genesis 2 should be understood as a transitional point of delineation between God Day Six and God Day Seven.[139] We have an especial affinity for the Hebraic "pre-rainy season" explanation of this text as posited by Collins.[140]

The Scriptures present God Day Seven as a Day that is comparatively unique to the other Days of creation. First, it is a Day of divine creative cessation and a celebration of completion. This Day signifies that God has finished the Original Relative Creation (a created order that, due to human-induced corruption, will eventually be brought to an end).[141] Second, it is a Day that has an open-ended chronological time duration (cf. Heb 4:1, 4, 9–10). The other Days have come and passed in *chronos*; each had a beginning and an end (note the "evening"/"morning" refrain). God Day Seven is still continuing (yet it is finite—see Matt 24:35). It will have a definitive ending (yet the time of that ending is still very unknown to humanity—see Matt 24:36–44). Third, the work of God continues in God Day Seven (cf. John 5:17—Jesus said, "My Father is working still, and I am working.") with a primary focus on the conservation of the present universe (Col 1:17), the redemption of humanity (John 3:16–17), and the completion of Final Absolute Creation (John 14:1–4; Rev 21:1–4[142]). The sustaining/redemp-

---

138. Ross holds to this view. (This evangelical view is as opposed to the higher critical view that proposes the Genesis text to be two independent creation stories interwoven together by a later redactor.)

139. The creation of Adam, and later Eve, is the relative boundary between God Day Six and God Day Seven in our correlation model.

140. Collins, *Genesis 1–4*, 121–24. He states (121): "In some particular but unnamed land, at the time of the year when the dry season was nearing its end and the rain cloud was rising, God 'formed' the man. Following the rest of the chapter, we find that after the Lord God placed man in the garden of Eden he fashioned a woman. Thus the pericope elaborates the very brief narrative of 1:27, declaring that what was presented as a single event in the first pericope actually was spread over a length of time. Only after he made the woman could God declare the whole thing 'very good' (1:31)."

141. This is as opposed to Final Absolute Creation, being prepared by God (John 14:1–4), which will never end.

142. John 14:1–4, [Jesus said] "Let not your hearts be troubled; believe in God, believe also in me. In my Father's house are many rooms; if it were not so, would I have told you that I go to prepare a place for you? And when I go and prepare a place for you, I will come again and take you to myself, that where I am you may be also. And you know the way where I am going."; Rev 21:1–4, "Then I saw a new heaven and a new earth; for the former heaven and earth had passed away, and the sea was no more. And I saw the holy city, new Jerusalem, coming down out of heaven from God, prepared as a bride adorned

tive/restorative work of God will continue until the Eschaton (cf. 2 Pet 3:9–13).[143] Essentially, in our model, the *seven* Days of the Creation Week represent the complete duration of chronological time (from the ex nihilo beginning of creation until the eschatological end of creation).[144]

The Advent and Placement of Humanity in Time

It is important now to discuss the advent and placement of humanity on the time-line of natural history. As previously asserted, our model places humanity a bit further back in time (1.9 Ma) than most creationists are willing to go. This placement comes from multiple considerations: the purpose of the genealogies of Genesis 5 and 11, the defining of humankind, and the timing of the Noahic Flood. These considerations will now be discussed.

The Genealogies of Genesis 5 and 11

First, we must consider the purpose of the genealogies of Genesis 5 and 11. Young-Earth Creationism uses these lists as an age tool to count backward in time from Abraham to Adam. It was by doing this that Bishop James Usshur reached his famous 4004 BC conclusion for the creation date of the world. Larry Pierce and Ken Ham concur with this thinking. They claim:

> Those who hold to the inerrancy of the Scriptures should reject all attempts to make the earth older than the Hebrew text warrants, which is about 4000 B.C. . . . The Scriptures themselves attest to the

for her husband; and I heard a loud voice from the throne saying, 'Behold, the dwelling of God is with men. He will dwell with them, and they shall be his people, and God himself shall be with them; and he will wipe away every tear from their eyes, and death shall be no more, neither shall there be mourning nor crying nor pain any more, for the former things have passed away."

143. 2 Pet 3:9–13, "The Lord is not slow about his promise as some count slowness, but is forebearing toward you, not wishing that any should perish, but that all should reach repentance. But the day of the Lord will come like a thief, and then the heavens will pass away with a loud noise, and the elements will be dissolved with fire, and the earth and the works that are upon it will be burned up. Since all things are thus to be dissolved, what sort of persons ought you to be in lives of holiness and godliness, waiting for and hastening the coming of the day of God, because of which the heavens will be kindled and dissolved, and the elements will melt with fire! But according to his promise we wait for new heavens and a new earth in which righteousness dwells."

144. See Mounce, *Expository Dictionary*, 637–38. In Hebriac thought, the number *seven* (Hebrew *seba*) represents "completeness."

fact that the secular dates given for the age of the universe, man's existence on the earth, and so on, are not correct, because they are based on the fallible assumptions of fallible humans. Nothing in observational science contradicts the time-line of history as recorded in the Bible... We can trust these genealogies because they are part of the infallible, inerrant Word of God.[145]

Unfortunately, the issue is not so easily settled. Does the Hebrew text really warrant such a view? The reality is that the matter is not simply about whether or not to *trust* the Genesis genealogies, but rather about how to appropriately *interpret* the genealogies. Are they given as a record of precise time (i.e., "chronogenealogies"), or are they given for some other reason?

For the purposes of this discussion, suffice it to say that we disagree with the Usshurian-type conclusion for the following reasons. First, despite the arguments of Pierce and Ham to the contrary, there are indeed discrepancies in the comparative genealogical orderings as recorded in Genesis and in Matthew. As Martin J. Wyngaarden says:

> [W]e cannot follow Ushur's chronology because the genealogies allow for gaps chronologically, even vast gaps... At any rate we are put on guard by the Scriptures of the O.T. and the N.T. themselves that the principle of omission obtains in these genealogies. In fact there may be very many missing links, besides Cainan, in Genesis 11:12 and Ahaziah, Joash, and Amaziah in Matthew 1:8. For the words beget and bear, like the words father and son, are used with more elasticity in Hebrew and Greek than in English.[146]

Such a view does not imply scriptural error. It simply gives credence to the notion that the genealogies are a part of the inspired Scriptures for reasons other than that of precise time records.

Second, the genealogical data seem to be placed in some form of evident systematic pattern. William Henry Green provides this arrangement analysis:

> The structure of the genealogies in Genesis 5 and 11 also favors the belief that they do not register all the names in these respective lines of descent. Their regularity seems to indicate intentional arrangement. Each genealogy includes ten names, Noah being the tenth from Adam, and Terah the tenth from Noah. And each ends with a father having three sons, as is likewise the case with

---

145. Pierce and Ham, "Are There Gaps in the Genesis Genealogies?," 181–82.
146. Wyngaarden, "Some of the Problems of Chronology in Genesis," 44.

The Correlation Model

the Cainite genealogy (4:17–22). The Sethite genealogy (chap. 5) culminates in its seventh member, Enoch, who "walked with God, and he was not, for God took him." The Cainite genealogy also culminates in its seventh member, Lamech, with his polygamy, bloody revenge, and boastful arrogance. The genealogy descending from Shem divides evenly at its fifth member, Peleg; and "in his days the earth divided." Now this adjustment in the genealogy in Matthew 1 into three periods of fourteen generations each is brought about by dropping the requisite number of names, it seems in the highest degree probable that the symmetry of these primitive genealogies is artificial rather than natural. It is much more likely that this definite number of names fitting into a regular scheme has been selected as sufficiently representing the periods to which they belong, than that all these striking numerical coincidences should have happened to occur in these successive instances.[147]

Green concludes "that the Scriptures furnish no data for a chronological computation prior to the life of Abraham; and that the Mosaic records do not fix and were not intended to fix the precise date either of the Flood or of the creation of the world."[148] This must be put into proper context.

Contrary to the typical ways of contemporary Western society, setting up genealogical patterns for emphatic purposes was a common practice in the culture of Near Eastern and Oriental antiquity. As John H. Walton says:

> Genealogies represent continuity and relationship. Often in the ancient Near East they are used for purposes of power and prestige. Linear genealogies start at point A (the creation of Adam and Eve, for example) and end at point B (Noah and the flood). Their intention is to bridge a gap between major events. Alternatively they can be vertical, tracing the descendants of a single family (Esau in Gen 36:1–5, 9–43). In the case of linear genealogies, the actual amount of time represented by these successive generations does not seem to be as important as the sense of completion or adherence to a purpose (such as the charge to be fertile and fill the earth). Vertical genealogies focus on establishing legitimacy for membership in the family or tribe (as in the case of the Levitical genealogies in Ezra 2).[149]

---

147. Green, "Primeval Chronology," 292.
148. Ibid.
149. Walton et al., "Genesis," 35.

Their conclusion is that "Genealogies are often formatted to suit a literary purpose" and not the purpose of a *modern* family tree.[150]

Following an in-depth numerical analysis of Genesis 5 and 11, involving both the Hebrew Masoretic text and the Greek Septuagint (both revealing, by the way, the same basic pattern), James L. Hayward and Donald E. Casebolt determined that "the probability that the Genesis age data represent a random distribution of age values is extremely low."[151] Though their research tool itself was not capable of determining exactly *why* the distribution of ages is the way it is, the tests did reveal the likelihood that the numerical structure of the genealogies was "biased"[152] (i.e., intentionally arranged). They further conclude:

> [T]he writer of the Genesis genealogies was more concerned with style than with chronology... the genealogies are neatly organized into two groups of ten patriarchs each, the first group containing antediluvians and the second group containing postdiluvians to the time of Abraham, the father of the Hebrew people... The Genesis genealogies would suit well a Hebrew writer's intent to show ancestral continuity between the Creator God and the Hebrew people, albeit in a somewhat stylistic fashion. But to force Genesis 5 and 11 to assume the role of "chronogenealogies" demands more of Scripture than we believe was intended by the inspired writers or is warranted by the evidence.[153]

We are in strong agreement with this view. The genealogies are indeed very important, but we believe have no use whatsoever in regard to the placement of humanity in time (or in determining the age of the Earth). That is simply not their purpose. Therefore, an Usshurian methodology is deemed to be inappropriate. We hold that God intended to communicate several important general realities: 1) the direct connection of historical humanity to the God of creation (Adam was created directly by God), 2) the direct connection of historical humanity to the Christ of redemption (the relationship of the Genesis genealogies with the Matthean [Matt 1:2–16, the royal descent of Jesus emphasizing his relationship to Abraham] and the Lukan [Luke 3:23–38, the royal descent of Jesus emphasizing his relationship to Adam] genealogies), 3) the direct connection of the historical Adamic fall

---

150. Ibid.
151. Hayward and Casebolt, "Genealogies of Genesis 5 and 11," 80.
152. Ibid.
153. Hayward and Casebolt, "Reactions," 8.

to a consequentially decreasing human life span (post-flood),[154] and 4) the keystone significance of the Noahic Flood in human history (note the equal number of presented generations antediluvian and postdiluvian signifying the *centrality* of the event in the pre-Abrahamic world). These purposes are all well served by the scriptural text. An absolute chronogenealogical interpretation serves only to detract from these purposes. Instead, we suggest that biblical references to time prior to the record of Abraham should be understood to be relative time rather than absolute time. With this in mind, we believe that any particular quest to discover the age of the Earth, the time for the origin of humanity, and the time of the flood must be made primarily from outside the pages of Scripture.

The Defining of Humankind

There is also the consideration of the defining of humankind. According to Ralph Stearley, "The fossil evidence for a long time-depth for human or human-like upright anthropoid primates is now considerable."[155] He asserts that the fossil record places hominids at least as far back as 6.5 million years with artifacts indicative of motor cognition and social organization.[156] Stearley's statement brings to bear a very important matter: the need for distinguishing between *human* and *human-like* anthropoid primates. Anthropology has delineated several taxonomic lineages, including that of the australopithecine genera (e.g., *Australopithecus, Paranthropus, Ardipithecus,* and *Kenyanthropus*) and the *Homo* genera (e.g., *Homo habilis, Homo rudolfensis, Homo erectus, Homo ergaster, Homo heidelbergensis, Homo neaderthalensis,* and *Homo sapiens*). We must attempt to reasonably discern just what is truly what in this seemingly multitudinous glob of primate types. This will be done systematically using typical descriptive criteria. We begin by presenting a description of the australopithecines:

---

154. Green, "Primeval Chronology," 290. Green states: "They [the given ages of people in the chronologies] merely afford us a conspectus of individual lives. And for this reason doubtless they are recorded. They exhibit in these selected examples the original term of human life. They show what it was in the ages before the Flood. They show how it was afterwards gradually narrowed down. But in order to do this it was not necessary that every individual should be named in the line from Adam to Noah and from Noah to Abraham, nor anything approaching it. A series of specimen lives, with the appropriate numbers attached, this is all that has been furnished us."

155. Stearley, "Assessing Evidences," 154.

156. Ibid.

> A major group of early hominids are the australopithecines, known only from the African continent, and including several species subsumed under three to five genera; *Australopithecus, Paranthropus, Ardipithecus,* and *Kenyanthropus* are probably stable. While australopithecine pelvises and limbs clearly indicate bipedality, particular features such as curved phalanges are interpreted as evidence for some arboreality, or alternatively, as evolutionary holdovers from arboreal ancestors. Australopithecines have a cranial capacity of around 400–450 cc. They are markedly sexually dimorphic. Australopithecines have been dated back to approximately 5.5 million years before the present (MYBP); incomplete and poorly understood remains occupy time horizons before that benchmark.[157]

Australopithecines (lit., "southern ape") were of ancient African origination dating from at least 5.5 million years ago. They were probably both bipedal and arboreal[158] and had very small cranial capacity. Further, australopithecines showed a very clear dimorphic differentiation between genders. Interestingly, this sexual dimorphism, particularly regarding physical size (with males usually being the larger), is very common in ape primates. Among some baboon species, for instance, the males are twice the size of the females.[159] This is quite impactful when one is trying to decide whether an australopithecine should be considered to be on the ape end or the human end of the primate spectrum.

Next, we present a description of the *Homo* genera, which we will place into "lower" and "higher" categories. First, the "lower" *Homo* genera:

> Hominids assigned to the genus *Homo* are known from stratigraphic horizons dating to about 2.5 MYBP. The earliest forms, [much] like the australopithecines, are known from Africa only and, at present, are assigned to *H. habilis* and *H. rudolfensis*. These earliest representatives of *Homo* are much shorter than modern *Homo sapiens* but possess a somewhat larger mean cranial capacity than that of the australopithecines, averaging 640 cc (range 590 to about 700 cc). *H. habilis*, while bipedal, possess feet which are rotated inward, such that locomotion on the hind limbs would

---

157. Ibid.

158. *Arboreal* is the propensity toward living in trees or having a dominant tree-climbing lifestyle.

159. Lindenfors and Tullberg, "Phylogenetic Analyses," 414.

The Correlation Model

have been extremely "pigeon-toed" and, in fact, not well suited for striding.[160]

We now present a description of the "higher" *Homo* lineage:

> Larger-statured representatives of the genus *Homo*, assigned to *H. erectus* (Asia) and *H. ergaster* (Africa) appear in the stratigraphic record about 2 MYBP. These forms approximate the height of modern humans, have body proportions (e.g., shape of pelvis, rib cage, projecting nose) which approximate those of modern humans, and have cranial capacities of between 900 and 1150 cc. The labyrinth of the inner ear attests to identical balancing ability while striding as that of modern humans. Females become relatively larger; sexual dimorphism is greatly reduced.[161]

As is obvious from these descriptions, there is a significant differentiation, not only between the australopithecines and *Homo*, but between the "higher" *Homo* genera and that of all the rest. This is particularly definitive in the areas of body proportion, bipedality, and especially cranial capacity. As anthropologist Richard E. Leakey said:

> [I]f one considers the brains of these ancient hominids one can see what might have been the differentiating factor between *Homo* and the australopithecines: intellect. The fossil skulls of the australopithecines suggest that there was little or no change in size or shape of their brains over a period of more than one million years. On the other hand, one of the outstanding characteristics of the early *Homo* species was the increase in the size of the brain.[162]

So, how should this web of data be interpreted? What are we to make of the postulations of modern anthropology? Can clarity among the taxonomic groupings somehow be found? Pat Shipman comments:

> Self-centeredly, human beings have always taken an exceptional interest in their origins. Each discovery of a new species of hominid—both our human ancestors and the near-relatives arising after the split from the gorilla-chimp lineage—is reported with great fanfare, even though the First Hominid remains elusive. We hope that, when our earliest ancestor is finally captured, it will reveal the fundamental adaptations that make us *us*.[163]

---

160. Stearley, "Assessing Evidences," 154.
161. Ibid.
162. Leakey, *The Making of Mankind*, 75.
163. Shipman, "Hunting the First Hominid," 25.

Shipman adds:

> The situation is deliciously complex and confusing . . . It is also humbling. We thought we could even identify a man in an ape suit or an ape in a tuxedo for what they were. Humans have long prided themselves on being very different from apes—but pride goeth before a fall. In this case, embarrassingly, we can't tell the ape from the hominid . . . Paleoanthropologists must seriously consider the defining attributes of apes and hominids while we wait for new fossils.[164]

In Shipman's commentary, notice the unclear delineation between between apes, hominids, and human beings. This is because popular evolutionary anthropology places humanity into the general category of hominid (with apes, of course, being the direct ancestors of hominids). The terms *primate*, *hominid*, and *human*, even *ape*, are frequently used in a blur of gray.[165] This is the nature of the philosophy of evolutionary continuity. From an evangelical Christian perspective, we sternly refute such a notion. In our model, we make a clear distinction between hominid and human. We certainly acknowledge the former existence of variant hominid taxa, but not with a direct ancestral connection to humanity. One perhaps should seriously consider the same ponderings as that of paleoanthropologist Peter Andrews, who—though an evolutionist—makes this astute comment: "[T]here was a whole proliferation of these apes . . . Why do they have to be ancestral to us?"[166] It is a very good question.

Without the evolutionary assumptions, it is indeed necessary to discern "the defining attributes" that separate apes and human-like hominids from true humanity in determining the placement of humans in time. In order to do this, we make two recommendations that go against the majority view of modern science, yet not without renowned scientific support. First, in concurrence with anthropologists Bernard Wood and Mark

---

164. Ibid.

165. Note the definitions of *hominid* taken from some popular dictionaries: "A primate of the family Hominidae, of which *Homo sapiens* is the only extant species" (*The American Heritage Dictionary of the English Language*, Houghton Mifflin, 2009); "Any primate of the family *Hominidae*, which includes modern man (*Homo sapiens*) and the extinct precursors of man" (*Collins English Dictionary*, HarperCollins Publishers, 2005); and "Any of various primates of the family Hominidae, whose only living members are modern humans" (*The American Heritage Science Dictionary*, Houghton Mifflin Company, 2005).

166. Gibbons, "In Search of the First Hominids," 1219.

The Correlation Model

Collard, we recommend removing *Homo habilis* from the *Homo* lineage and placing it into the australopithecine lineage. Wood and Collard state:

> A general problem in biology is how to incorporate information about evolutionary history and adaptation into taxonomy. The problem is exemplified in attempts to define our own genus, *Homo* . . . conventional criteria for allocating fossil species to *Homo* are reviewed and are found to be either inappropriate or inoperable. We present a revised definition, based on verifiable criteria, for *Homo* and conclude that two species, *Homo habilis* and *Homo rudolfensis*, do not belong in the genus. The earliest taxon to satisfy the criteria is *Homo ergaster*, or early African *Homo erectus*, which currently appears in the fossil record at about 1.9 million years ago.[167]

Their solution and revised definition is as follows:

> We suggest that a fossil species should be included in *Homo* only if it can be demonstrated that it (i) is more closely related to *H. sapiens* than it is to the australopiths, (ii) has an estimated body mass that is more similar to that of *H. sapiens* than to that of the australopiths, (iii) has reconstructed body proportions that match those of *H. sapiens* more closely than those of the australopiths, (iv) has a postcranial skeleton whose functional morphology is consistent with modern human-like obligate bipedalism and limited facility for climbing, (v) is equipped with teeth and jaws that are more similar in terms of relative size to those of modern humans than to those of the australopiths, and (vi) shows evidence for a modern human-like extended period of growth and development.[168]

In other words, Wood and Collard are attempting to set up a systematic way of discerning a differentiation between true humans (i.e., the "human genus" = those like *Homo sapiens*) from that of non-humans (those like the australopithecines).[169] In essence, they hold that a species should be considered to be *Homo sapiens* if it is more like *Homo sapiens* than australopithecine; likewise, a species should be considered to be australopithecine if it

---

167. Wood and Collard, "Human Genus," 65.
168. Ibid., 70–71.
169. Many anthropologists, due to evolutionary assumptions, prefer to describe australopithecines as proto-human or proto-hominid (as opposed to non-human). For instance, see Leakey and Lewin, *Origins Reconsidered*, 189, 191; see also, Wallbank et al., *Civilization: Past & Present*, 16.

is more like australopithecine than *Homo sapiens*. It is their opinion that *Homo habilis* (and *Homo rudolfensis*) does not fit the above *Homo sapiens* criteria and thus should be relegated to australopithecine status. Furthermore, they believe that the earliest species to fit the criteria is *Homo erectus* (and *Homo ergaster*, which is the early African version of *Homo erectus*).

Second, in concurrence with anthropologist Milford H. Wolpoff (as well as Wood and Collard), we recommend the merger of *Homo erectus* into the *Homo sapiens* line. Wolpoff states:

> We propose here to merge *Homo erectus* within the evolutionary species *Homo sapiens*. The origin of *Homo erectus* lies in a cladogenic event at least 2.0 myr ago. We view the subsequent lineage as culturally and physically adapted to an increasingly broad range of ecologies, ultimately leading to its spread across the old world prior to the beginning of the Middle Pleistocene. *Homo erectus* differs from *Homo habilis* in a number of ways. The vast majority of these distinctions also characterize *Homo sapiens*. The few distinctions of *Homo sapiens* that are not shared with *Homo erectus* appear to be responses to, or reflections of, continuing evolutionary trends of increasing cultural complexity, increasing brain size, and the progressive substitution of technology for biology.[170]

Wolpoff elaborates further:

> *Homo erectus* is a polytypic species, divided into several distinct geographic variants which each show at least some genetic continuity with the geographic variants of the polytypic species *Homo sapiens* that is reflected in shared combinations of morphological features. There is no distinct boundary between *Homo erectus* and *Homo sapiens* in time or space, and cladogenesis does not seem to mark the origin of *Homo sapiens*. Instead, the characteristics of *Homo erectus* and *Homo sapiens* are found to be mixed in seemingly transitional samples from the later Middle Pleistocene of every region where there are human remains. The regional ancestry of *Homo sapiens* populations makes monophyly impossible for the species if the earlier populations are in a different species. We interpret this to mean that there is no speciation involved in the emergence of *Homo sapiens* from *Homo erectus*. These reasons combine to require that the lineage be regarded as a single evolutionary species.[171]

---

170. Wolpoff et al., "Case for Sinking *Homo Erectus*," 341.
171. Ibid.

The Correlation Model

Note Wolpoff's key assertions: 1) "There is no distinct boundary between *Homo erectus* and *Homo sapiens*," 2) "cladogenesis does not seem to mark the origin of *Homo sapiens*," and 3) "there is no speciation involved in the emergence of *Homo sapiens* from *Homo erectus*." Wolpoff is positing that *Homo erectus* and *Homo sapiens* should be considered to be of the same lineage and "a single evolutionary species." To borrow Wood and Collard's description, this is the *human genus*.

In effect, the adherence to these recommendations draws a definitive boundary line in the sands of chronological time. In our model (non-evolutionary), it pushes *Homo habilis* into the category of non-human hominid and places *Homo erectus* into the category of true human being. This also means that the original appearance of modern humanity is placed at least to about the 1.9 million year point (and possibly even a bit earlier).[172]

The Timing of the Noahic Flood

Finally, there is the consideration of the timing of the Noahic Flood. If the flood was global in scope, which we believe to be the case, then it is our determination that there is a probable cause and effect relationship between the Deluge (cause) and the Pleistocene Ice Age (effect). We conclude this based on a modified Old-Earth application of the Michael J. Oard Flood-Ice Age Model.[173] Oard, a meteorologist, posits that a global flood would result in conditions very favorable for worldwide glaciation within 200 to 1,700 years post-flood (with about 500 years most probable).[174] This means that, since the Noahic Flood was an act of divine judgment upon sinful humanity (Gen 6), the human race must have existed prior to the Pleistocene Ice Age (definitive glacial-interglacial cycles were initially manifested by approximately 1.81 Ma with extremely dominant cycles beginning at 0.6 Ma[175]). This also smoothly fits the time line of 1.9 million years ago for

---

172. Note the slight relative time differential: Wood and Collard assert 1.9 Ma; Wolpoff asserts 2.0 Ma.

173. Oard, "A Post-Flood Ice-Age Model," 8–26.

174. Ibid., 7. See also, Clark, *Fossils, Flood, and Fire*, 180. He seems to favor Ice Age glaciation about 1000 years post-flood.

175. McGowan et al., "Neogene and Quaternary coexisting," 249–62. McGowan states, "The [new] Pliocene/Pleistocene boundary is not an extreme climactic event but even so is well marked. The strongest change comes with the mid-Pleistocene transition to the ~100 kyr—dominated and strongly amplified glacial-interglacial oscillations occur from 0.6 Ma, with four glacials outstanding: MIS 2, 6, 12, and 16" (248).

the advent of true humanity in accordance with the postulations of Wood, Collard, and Wolpoff. Therefore, in this time placement paradigm, Adam would have existed *at least* 90,000 years (possibly much earlier) before the Noahic catastrophe (which, applying the Oard model, would have occurred 200 to 1,700 years before the 1.81 Ma boundary). This time frame gives the historic Adam and Eve a significant opportunity to live and subsist in the pristine Edenic world (Gen 2:8–25) prior to the fall and subsequent advent of death and corruption (Gen 3); it also provides ample time (post-fall) for human sin to manifest itself into the extremely wicked pre-flood conditions as recorded in Genesis 6:1–5.

Some have suggested that difficulties of compatibility exist between the belief in a historical Adam and the holdings of modern anthropology. For example, Paul Seely (a notable non-concordantist) presents the following assertion and proposed solution:

> The perspectives of 20th century anthropology are incompatible with the acceptance of the literal historicity of Genesis 2 and 3. Anthropology's first man must be dated before Neolithic times; the literal man of Genesis 2 and 3 must be dated in Neolithic times. The legitimate use of anthropology resolves the conflict by leading to the recognition that Adam is a figurative person, who harmonizes with both anthropology and biblical theology.[176]

According to Seely, the best way to deal with the alleged difficulty is to recognize that the biblical Adam was not a historical individual, but rather a figurative title given for representative humanity. Such concerns, as that expressed by Seely, are certainly well taken, but unnecessary. First, they are in response to an Usshurian-type chronology (namely, Adam was created sometime between 6 to 10 ka, forcing him into a Neolithic time-frame), which we have already deemed to be inappropriate. In our model, from the perspective of *time*, Adam (c. 1.9 Ma) most certainly was Paleolithic (2.6 Ma to 12,000 ka—"Early Stone Age") and not Neolithic; yet, from the perspective of *culture* and *technology*, we believe that Adam was not primitive (as the traditional understanding of Paleolithic culture posits). Second, it is also based on an understanding of human culturo-technological advancement that must be called into question on scriptural grounds.

---

176. Seely, "Adam and Anthropology," 88.

The Matter of Adamic Culture and Technology

Using a "year" as the symbolic equivalent to the totality of geological time, T. Walker Wallbank presents the typical evolutionary view of human technological development:

> In this "year" members of *Homo erectus*, our ancient predecessors, would mount the global stage only between 10 and 11 p.m. on December 31. And how has the human species spent that brief allotment? Most of it . . . has been given over to making tools out of stone. The revolutionary changeover from a food-hunting nomad to a farmer who raised grain and domesticated animals would occur in the last sixty seconds. And into that final minute would be crowded all of humanity's other accomplishments: the use of metal, the creation of civilizations, the mastery of the oceans, the harnessing of steam, then gas, electricity, oil, and, finally, in our lifetime, atomic energy. And now humankind's technological genius has enabled it to escape from age-old bondage to the earth and to initiate an interplanetary age.[177]

Modern anthropology divides human prehistory into culturo-technological *stages*. They are known as the Paleolithic, the Mesolithic, and the Neolithic.[178] These are the three sub-stages of the "Stone Age."[179] To put things in larger perspective, science posits that the Stone Age was followed successively by the Copper Age, the Bronze Age, and the Iron Age. The basic idea as set forth by evolutionary anthropology is that humanity has gradually moved from being extremely primitive (using the crudest of stone tools to accomplish rudimentary tasks) to developing and implementing more and more complex metal tools and accomplishing much more difficult tasks. Thus, in this view, human technology moved from being very primitive and simple to being very modern over the course of the last 2.6 million years (this time duration, of course, includes *Homo habilis* in the human lineage—which we refute). Further, this also means that evolutionary anthropology holds that the majority portion of human history was

---

177. Wallbank et al., *Civilization*, 1.
178. Ibid., 18–22.
179. Ibid., 19, 21. Note that, while these terms—Paleolithic, Mesolithic, and Neolithic—are sometimes equated to some degree with chronological time, they are actually cultural and technological stages independent of strict chronological time. They often, in fact, overlap.

Stone Age. However, the question must be asked: Is this the best interpretation of the evidence available, especially in light of the scriptural record?

Donald E. Chittick claims that it is not the best interpretation and presents a different view. He asserts:

> Even though the early people of the earth, Adam and Eve and their descendants, were in a fallen and degenerating state, they were not "primitive" people. Being closer to Creation than we are, they had not degenerated as far as we have. Their bodies were far more perfect, their minds more alert and capable, and their lifetimes longer. With their good health, keen observational powers, and alert minds, they soon began to develop a high level of science and technology. It is an error to assume, as current culture does, that early man was not mentally highly capable, or that he was ignorant of what we would call science and scientific principles.[180]

In light of the postulations of modern anthropology, can the Chittick view be empirically justified? We believe that it can. It seems that God may have left just a few traces of physical evidence to warrant contemporary humanity giving the Genesis Adamic record a very good look. The key is to make sure that the empirical data is interpreted within the framework of the scriptural revelation. Thus, before we look at the empirical data, we will examine a few passages from Genesis, namely from Genesis 1–4, that provide clues as to Adamic culture and technology. It is important to keep in mind that these texts all record descriptions of *antediluvian* (pre-flood) humanity.

The Clues about Ancient Adamic Culturo-Technology from Scripture

The scriptural revelation provides several important clues as to the advancement level of early humanity. It shows that Adam and his immediate progeny were capable of some very functional degree of verbal communication, tent making, domesticated animal husbandry, advanced music, metallurgy, etc. In other words, Adamic humanity was intelligent and lived in a civilized and ordered society.[181] Rene Noorbergen affirms this thought:

---

180. Chittick, *Puzzle of Ancient Man*, 41.
181. Ibid., 41–45.

The Correlation Model

According to the Genesis account, the antediluvian people were highly knowledgeable, being the first to develop agriculture, animal husbandry, construction, architecture, political organization, metal working, the abstract arts, mathematics, chronology and astronomy. What's more, while the Genesis record tells us that there were altogether only ten generations of antediluvians, they developed the majority, if not all, of the basic elements of civilization by the sixth generation. Now if the antediluvians, beginning with the raw earth, were able to master the arts of civilization in the first six generations of their existence, we may well wonder to what degree they further developed and refined those arts in the remaining four prior to the Deluge.[182]

We now provide a concise, yet descriptive breakdown of what the Genesis text (quite non-exhaustively, by the way) communicates about early humanity.

First, Adam engaged in complex verbal communication. In Genesis 1:28–30, God speaks directly to Adam and Eve giving them the creation and dominion mandate. In Genesis 2:16–17, God likewise spoke to Adam concerning what he could and could not eat. Genesis 3 presents two significant verbal encounters: in Genesis 3:1–7, Eve encounters the serpent and engages in a "fatal" conversation; Genesis 3:8–13 presents the encounter of God walking in the Garden to visit with Adam and Eve culminating in a verbal exchange. These texts imply the Adamic capacity for verbal communication, developed cognitive ability, and deep personal relationships (including, notably, spiritual relationships).

Second, Adam's descendants engaged in tent making and the raising of domestic animals. Genesis 5:20 says that Jabal "was the father of those who dwell in tents and have cattle." It is possible that this also could be making reference to early shepherds.

Third, Adam's descendants engaged in advanced music. Genesis 5:21 says that Jubal "was the father of all those who play the lyre and the pipe." The specific instruments mentioned are rather complex and require special musical ability.

Fourth, Adam's descendants also practiced metallurgy. Genesis 5:22 states that Tubal-Cain "was the forger of all instruments of bronze and iron." Such a level of technology—the forging of metal—is very advanced.

---

182. Noorbergen, *Secrets of the Lost Races*, 3–4.

Is the Genesis text fallacious? It should be noted that modern anthropology has no real regard for the Scriptures in the search for truth. Wood, for example, comments:

> With very few exceptions Western philosophers living in and near the Dark Ages (5th to 12th centuries) supported a biblical explanation for human origins. This changed with the rediscovery and rapid growth of natural philosophy that was only later called science . . . The move away from religious dogma was especially important for those who were interested in what we now call the natural sciences, such as biology and the earth sciences.[183]

Wood seems to be subtly implying that a biblical view of human origins is merely a form of religious dogma and something that should be relegated to the "Dark Ages" of human thinking. At best, the Scriptures are viewed as a primeval account that simply shows the pre-scientific attempts of people to explain humanity's beginning. Thus, with this view (which is, by the way, quite evolutionary), early humanity can be easily understood as having been very primitive and backward. However, if the Bible is understood to be historically and scientifically veracious within its God-ordained purpose, a very different picture of early humanity is painted entirely.

The Clues about Ancient Adamic Culturo-Technology from Empirical Observation

There are large and seemingly inexplicable empirical clues that cast big question marks on the conclusions of orthodox science as to the backwardness of early humanity. The first clue involves human artifacts. Artifacts are physical evidence that provide useful information about early humanity. Much like Eerdman's earlier conclusion about fossils in rocks, ancient human artifacts serve as an unbiased record that must somehow be reasonably interpreted. They are simply what they are. As to meaning, human artifacts, unless tampered with, should be accurately reflective of the reality of past human existence.[184] Chittick makes a remarkable observation:

---

183. Wood, *Human Evolution*, 9. Interestingly, Wood equates "natural *philosophy*" with what "was only later called [natural] *science*." This implies that all disciplines, including science (and theology), are permeated with a degree of subjectivity and presupposition that can directly affect the arrival of investigative conclusions.

184. Chittick, "The Puzzle," 7–8.

The Correlation Model

> [S]uch artifacts pose a very great problem for the evolutionary picture that man developed upward from the animals. If evolutionism is correct and man did actually evolve up from the animal, then human artifacts should reflect that fact. The more ancient an artifact, the more "primitive" it ought to be. The actual case with the evidence, however, seems to be just the opposite... In fact, the cultural remains of ancient man are so at odds with the evolutionary picture for man's origin that a special term has been coined to describe these artifacts. The term is OOPARTS, an acronym for Out Of Place Artifacts.[185]

The implication of Chittick's statement is significant. There are advanced human artifacts in places where they should not be according to orthodox science. Just a few examples of OOPARTS include the following: a machine-made metal cube comprised of a steel-and-nickel alloy found in Tertiary coal ("Salzburg Cube"), a unique electrical device—similar to a spark plug—found inside sedimentary rock dated at one million years old ("Coso Artifact"), a gold chain found in a vein of Carboniferous rock ("Culp Chain"), and later generation technology such as the objects resembling electric batteries found in Iraq and Iran amidst Babylonian and Medo-Persian ruins as well as evidence of the possible use of electricity in ancient Egypt.[186] The typical attempts to answer these OOPARTS are to interpret them as either strange objects of nature or as contemporary technology that somehow found itself inserted into strange locations. Regardless, there are many more such artifacts that have been found and continue to be found. They are not simply an anomaly that can validly be ignored or nominally explained away.[187] With openness to a new set of presuppositions, many of these seemingly outlandish notions can be understood much more as legitimate possibilities. As Noorbergen states:

---

185. Ibid., 8.

186. Noorbergen, *Secrets of the Lost Races*, 40–62. See also Cremo and Thompson, *Forbidden Archaeology*. In this controversial, yet massive and well-documented work, they present a very comprehensive case for the high culturo-technological advancement of human society in the extremely antiquated (Ma) past. See especially 795–813. It should be noted that Cremo and the late Thompson represent Hindu creationism (Vedic tradition) backgrounds.

187. Chittick, *Puzzle of Ancient Man*, 12. He states: "At first, one might be tempted to think that these and other 'out-of-place' artifacts are just a few oddities. However, many such artifacts are known and are documented in various sources. Material in these references shows that OOPARTS are not an anomaly."

> The continual discovery of various OOPARTS is beginning to bring us closer to our forefathers in a way we never imagined. Studying the various remnants of their technological achievements has added an entirely new dimension to our concept of the lives of these peoples. We are now beginning to learn how they may have lived, and scattered fragments of their high technology are providing us with little hints that enable us to effect an imaginary reconstruction of some of their accomplishments.[188]

The discovery of OOPARTS implies the potentiality that early humanity (both antediluvian and, by extension, post-flood Noahic) was not primitive, but rather quite advanced in culture and technology. This perspective provides the keen researcher with just a glimpse of empirical data that potentially corroborates the aforementioned descriptions of early humanity as found in the Genesis text. If Adam was created and lived 1.9 million years ago (or even earlier) as we suggest in our correlation model, then OOPARTS are all the culturo-technological remains that *should* be found. From a chronological perspective, the Noahic Flood followed the creation of Adam. As an act of divine judgment upon a terribly depraved humanity, the Flood virtually blotted antediluvian culture and technology from the annals of human history. The Ice Age, following the flood, also must have had profound effect upon human culturo-technology as well. Will Durant, a renowned historian, makes a strong comment that can be very indicting of orthodox anthropology: "These primitive cultures [namely, Sumer, Egypt, Babylon, etc.] . . . were not necessarily the [earliest] ancestors of our own; for all that we know they may be the degenerate remnants of higher cultures that decayed when human leadership moved in the wake of the receding ice from the tropics to the north temperate zone."[189] Durant's point is that modern human society may actually be the descendant of unknown peoples who were more advanced than the earlier civilizations that are known to us—maybe even as culturally and technologically advanced as 21st century humanity. According to Durant:

> Obviously historiography cannot be a science. It can only be an industry, an art, and a philosophy—an industry by ferreting out the facts, an art by establishing a meaningful order in the chaos of materials, a philosophy by seeking perspective and enlightenment . . . We do not know the whole of man's history; there were probably

---

188. Noorbergen, *Treasures of the Lost Races*, 3.
189. Durant, *Our Oriental Heritage*, 90.

The Correlation Model

many civilizations before the Sumerian or the Egyptian; we have just begun to dig! We must operate with partial knowledge, and be provisionally content with probabilities; in history, as in science and politics, relativity rules, and all formulas should be suspect.[190]

Note the comment, "there were probably many civilizations before the Sumerian or the Egyptian." The scope of this thought includes not only the possibility, but perhaps the probability, of "lost" societies that once existed after the Noahic Flood and those that existed long *before* the flood. Perhaps contemporary human arrogance remains our greatest detriment.

The only remaining signs of the existence of antediluvian society would be the scriptural record, some extant human fossils (e.g., *Homo erectus* remains, etc.), perhaps the occasional yet growing number of OOPARTS, and any other lasting results of the "memory" that Noah and his family carried with them over the diluvial divide (though specific associations probably cannot be made today). In effect, after the flood, humanity had to start over. Yet, this time, fallen humanity, being much further from original creation, did not begin with the same pristine capacity as did Adam. Post-diluvian humanity, though intelligent and having some degree of culturo-technological "memory" from the past generations, had to largely struggle from scratch to rebuild another version of a society almost completely eliminated by God. Thus, modern anthropology, having no place for the Noahic Flood (or the later Babelian dispersion) in its schematic, considers only the very sparse remnants of antediluvian human existence jumbled up with extant portions of ancient post-diluvian society without any reference to the scriptural record. This method of interpreting truth leaves wide evidential gaps in its attempt to explain phenomena.

The second empirical clue as to the extreme advancement of early humanity involves the possibility of ancient cartography and global exploration. This is an extremely controversial, yet fascinating area of study. Chittick states:

> An increasingly large amount of evidence indicates that early post-Flood dwellers did indeed possess an amazing amount of accurate knowledge about the globe. One indication of that is the existence of copies of ancient maps such as the Piri Reis Map . . . The Piri Reis Map and other copies of very ancient maps show that the ancients not only were aware of world geography, but they actually mapped it as well . . . There are numerous other ancient

---

190. Durant and Durant, *Lessons of History*, 12–13.

maps which have been discovered during the last 100 years. It is the accuracy of these ancient maps that amazes many moderns. Accurate map making requires a high degree of technical skill as well as considerable knowledge of mathematics.[191]

The Piri Reis Map is dated 1513. However, it embodies certain cartographical data that some scholars believe could only be acquired from measuring techniques that were not available to modern society until the 19th century.[192] A truly astounding reality is that the Piri Reis Map is not an original. It is known to be the composite of at least twenty much older maps, some dating possibly many centuries or millennia into the past (prior to 1513).[193]

The Piri Reis Map has several extraordinary features. First, the map (lost in the mid-1500's and not rediscovered until 1929[194]) shows the continents of Africa and South America in correct relative longitude with one another. However, in the 1500's, longitude was determined somewhat by guesswork. It was not until nearly 400 years later that modern cartography (using advanced technology) determined the correct longitudinal relationship of the two continents (revealing that the Piri Reis Map was fairly correct in its calculations). Second, although disputed, the Piri Reis Map of 1513 possibly shows the continent of Antarctica.[195] This is important because the existence of Antarctica, although theorized and seemingly noted on a number of medieval cartographical representations (other than the Piri

---

191. Chittick, *Puzzle of Ancient Man*, 58–59.

192. Noorbergen, *Secrets of the Lost Races*, 98. See also, Hapgood, *Maps of the Ancient Sea Kings*, 41–49.

193. Hapgood, *Maps*, 39. He concludes: "[T]he Piri Reis Map of 1513 ... represent[s] originally separate source maps of smaller areas, which appear to have been combined in a general map by the Greek geographers of the School of Alexandria. ... [I]n most cases the errors on the Piri Reis map are due to mistakes in the compilation of the world map, presumably in Alexandrian times, since it appears ... that Piri Reis could not have put them together at all. The component maps, coming from a far greater antiquity, were far more accurate. The Piri Reis Map appears, therefore, to be evidence of a decline of science from remote antiquity to classical times (39)."

194. Stiebing, Jr., *Ancient Astronauts*, 91.

195. There are those who disagree that the "southern landmass" on the Piri Reis Map is actually intended to be what we know today as Antarctica. See Cuoghi, "The Mysteries of the Piri Reis Map," who claims "that it represents nothing more than the extremity of the south-american continent, an approximate representation made possible by means available at that time." Also, see Dutch, "The Piri Reis Map," who claims that a comparison between what appears on the Piri Reis Map and Antarctica reveal only "vague similarities."

THE CORRELATION MODEL

Reis Map),[196] was not officially confirmed until 1820.[197] Furthermore, the "Antarctic" coastline as shown on the map is roughly similar (yet without the ice) to the modern delineation, although it does not show the waters of the Drake Passage that currently separates Antarctica from South America.[198] Due to the existence of the current levels of immensely deep polar glaciation (mean depth of 6,355 feet[199]), contemporary researchers have only been able to confirm the Antarctic cartography through advanced technologies such as seismic echo, ground-penetrating radar, and satellite imaging. What does this mean? According to Noorbergen:

> The unavoidable conclusion was that Piri Reis must have possessed charts drawn by someone who had mapped Antarctica *before* the ice cap covered the southern continent. The Piri Reis map could not have been a hoax, for no one in 1929 [the year it was rediscovered], let alone in 1513, could have reproduced the geographical knowledge this chart contained. . . . It [the Piri Reis Map] is the product of an unknown people antedating recognized history.[200]

In other words, in Noorbergen's thinking, the ancient unknown cartographers completed their Antarctic charting prior to its current polar glaciation.[201] This would, of course, require the existence of a very antiquated, yet

---

196. Hapgood, *Maps*, 79–112. For instance, these include the Oronteus Finaeus Map, the Hadji Ahmed Map, the Mercator Map, and the Schoner Globe.

197. Turney, *1912: The Year*, 3, 12.

198. Hapgood, *Maps*, 72, 83–87. He surmises that there is a very logical reason for the absence of the Drake Passage on the map: "By using an oversized map the compiler was forced to crowd the Antarctic Peninsula up against Cape Horn, squeezing out the Drake Passage almost entirely" (83). This inaccuracy goes back to "some ancient period, on a source map used by Piri Reis, of a large part of the coastline of South America" because "There was simply no room for it!" (87).

199. Fretwell, et al., "Bedmap2," 390.

200. Noorbergen, *Secrets of the Lost Races*, 96–97.

201. Is this truly realistic? Many scholars assert that Antarctic glaciation has existed in some form or fashion for 35–40 million years (since mid-Tertiary) [e.g., see Ingolfsson, "Quaternary glacial and climate history of Antarctica," 3], although others claim that ice core evidence may reasonably show its appearance to be much more recent in time (since late Quaternary) [e.g., see Vardiman, *Ice Cores and the Age of the Earth*, 31–32; 65–66]. Nevertheless, it should be noted that conventional science generally holds that Antarctica was completely covered in ice by 15 Ma and reached its current glacial depth and extension around 6 Ma. HOWEVER, it is interesting that a very large semi-*fossilized* temperate Antarctic forest relatively dated at 3 Ma has recently been discovered implying a definitive and significant break or waning in the ancient glaciation (see Stefi Weisburd, "A forest grows in Antarctica—an extensive forest may have flourished about 3 million

very technologically advanced human civilization. Noorbergen provides this summative comment:

> The only realistic conclusion one can reach on the basis of the accumulative evidence of the medieval maps [such as the Piri Reis Map] is that they all have their origin in source maps constructed by an advanced civilization antedating any of the known ancient cultures. Years before the Egyptian, Babylonian, Greek and Roman civilizations existed, at a time when the Antarctic and Arctic were just beginning to feel the advance of the unyielding sheets of glacial ice, this unknown culture possessed a knowledge of cartography comparable to what we have today. These people knew the correct size of the earth; they used spherical trigonometry in their mathematical measurements; and they utilized ultramodern cartographical projections. In addition to their science, these surveyors must have had at their disposal an advanced form of technology—instruments, and trained specialists to use them, for measuring longitude and latitude. The pre-ancient civilization of the past . . . must have been organized and directed on a global scale.[202]

In light of the possibilities of "pre-ancient" global cartography, there is a very interesting thought concerning the enigmatic individual mentioned in Genesis 10 named Peleg (who was a descendant of Noah through the lineage of Shem). Genesis 10:25 states, "the name of the one was Peleg, for in his days the earth was divided." According to Walton et al., "While this has traditionally been taken to refer to the division of the nations after the Tower of Babel incident (Gen 11:1–9), other possibilities exist. It could, for instance, refer to a division of human communities into sedentary farmers and pastoral nomads; or, possibly a migration of peoples is documented here that drastically transformed the culture of the ancient Near East—perhaps one represented in a break-off group traveling southeast in Genesis 11:2."[203] Noorbergen presents still another possibility: "[I]t could also mean division as in 'allotment, marking off an area, a measurement.' A more ac-

---

years ago.") This suggests that the current modern level and extent of glaciation is much more recent (3 Ma or just later) than has been conventionally surmised. It also makes plausible the idea that the gap of deglaciation (perhaps from before 3 Ma to beyond 2 Ma?) coincides with the advent of early humanity (c. 1.9 Ma or earlier in our model) and provides the open window by which (antediluvian?) cartographers could have possibly charted the southern continent (before the current re-glaciation occurred).

202. Noorbergen, *Secrets of the Lost Races*, 98–99.
203. Walton et al., "Genesis," 41.

curate translation of this historical passage could therefore read, 'Peleg, in his day was the earth measured, or surveyed.'"[204] If Noorbergen's view is taken, the Scriptures may just be providing a subtle hint at the culturo-technological advancement of early post-flood (yet pre-Ice Age) humanity. The point of this is not so much to present a contrast between the level of diluvian (both pre and post) technology versus that of contemporary technology (e.g., comparing the technical accuracy of the pre-Piri Reis cartography to contemporary cartography, etc.), but rather to suggest the existence of certain empirical trace evidences that stand in agreement with the biblical revelation (namely, the assertion of Genesis that Adamic antediluvian humanity was intelligent, civilized, and advanced).

There are additional possibilities as well that can provide clues as to the advancement level of very early human civilization. Without going into in-depth detail (and as strange as it may seem), suffice it to say that there are both material artifacts and extant ancient writings that potentially suggest that antediluvian humanity was no stranger to aerial flight and maybe even something akin to atomic power. While the interpretation of the artifacts and the degree of historicity for the ancient writings is rightly debated, the artifacts do include, for instance, the large and intricate designs found in the Nazca Dessert of South America[205] as well as individual objects bearing amazing resemblance to modern airships.[206] As to the ancient writings, they include frighteningly graphic descriptions of devastating wars among earlier peoples that certainly do not fit the common stereotypical picture of battles involving ape-people with rocks and spears.[207] Therefore, the various authors seem to be either the recorders of vivid interpretations of very early human history or the possessors of an extremely "modern"

---

204. Noorbergen, *Secrets of the Lost Races*, 99.

205. Chittick, *Puzzle of Ancient Man*, 181–84. Referring to the amazing designs located in the Nazca Dessert near Lima, Peru, Chittick states: "The fact that the designs in the Nazca Desert can only be viewed effectively from the air should at least suggest the possibility that the constructors of the Nazca designs may have possessed the capacity for flight. And yet modern secular archaeologists who believed man had evolved upward from a more primitive state were forced to deny the obvious. These modern archaeologists could not believe that people back then might have had air travel (184)."

206. Ibid., 187.

207. For instance, see the Indian Sanskrit epics known as the *Mahabharata* and the *Ramayana*.

imagination. Either way, a competent researcher must be willing to reason and wonder.[208]

It is interesting to look at the first possibility (i.e., that we possess records of vivid interpretations of early human history, particularly as related to ancient wars) in the context of the pre-flood passage of Genesis 6:11–13:

> Now the earth was corrupt in God's sight, and the earth was filled with violence. And God saw the earth, and behold, it was corrupt; for all flesh had corrupted their way upon the earth. And God said to Noah, "I have determined to make an end of all flesh; for the earth is filled with violence through them; behold, I will destroy them with the earth."

Very significantly, the above text does not merely speak of the general corruption of antediluvian humanity, but specifically states twice that the earth was "filled with violence." The scriptures poignantly proclaim that the manifestation of the human *corruption* in that day was specifically that of human *violence*. Thus the notion can plausibly be conceived that fallen antediluvian humanity had grown so technologically advanced on a global scale, yet so corrupt in the usage of the technology, that divine judgment was initiated for the ultimate preservation of the human race.

It is our contention—based primarily on the biblical revelation, and secondarily on trace empirical data—that the time of human existence has witnessed at least three "peaks" of advanced human civilization. First, there was the antediluvian society that was concluded by the Noahic Flood. Second, there was the immediate post-diluvian society that was in a sense concluded by the Babelian Dispersion. Finally, there is the current human civilization that exists in anticipation of its conclusion at the Eschaton.

---

208. See Schmidt and Frank, "The Silurian Hypothesis," 1–9. In a cooperative venture by scientists from NASA and the University of Rochester, an ongoing investigation is now being made as to the possibility that other long, lost industrialized civilizations, possibly even pre-Quaternary, existed on Earth prior to our contemporary civilization. In this paper, they propose tests for the detection of potential geological fingerprints with a special focus on possible ancient industrial energy signatures. (Their hypothesis also leaves open the possibility for an early civilization to be other than human.) *Note:* Our point is that modern science is beginning to entertain the thought that our current society may not be the first modern, industrialized, and high-tech civilization to ever exist on the planet Earth.

The Babelian Dispersion

Another prominent issue of anthropology is the matter of human distribution and racial diversity across the Earth. Scholars are divided into two primary modes of thought in this regard. There is the monocentric school and the polycentric school.[209]

The monocentric school posits that humanity appeared and evolved from a single line of ancestry. The origin of people either occurred in a specific and narrow area (e.g., Mesopotamia, or East and West Turkana, etc.) or in a specific and broader region (i.e., a region covering northern Africa, west Asia, and parts of central and south Asia). According to the monocentric school, the first people migrated from a singular and specific place of origin to other areas of the Earth resulting in the evolution of a diversity of distinct racial types in various geographic regions.[210] This thinking gives rise to theoretical concepts such as the "Out Of Africa" model[211] (which is the current majority view of modern anthropology). It is largely based on the fact that the earliest known humans have been found in Africa (dated at 1.9 Ma). Outside of Africa, the earliest known humans have been found at Dmanisi in the Causcasus (dated at 1.8 Ma), China and Indonesia (dated at 1.7 Ma), and Ubeidiya in Israel (dated at 1.5 Ma), respectively.[212] However, Wood wisely cautions researchers to be aware that the "absence of evidence is not evidence of absence." These known human remains are merely what anthropology has discovered thus far.[213]

The polycentric school posits that humanity appeared and evolved from at least four different ancestral lines in different parts of the world. These include the African, Mongoloid, Australian, and the Caucasoid (or Eurasian) lineages. According to the polycentric school, the process of human development through multiple and distinct lineages occurred independently of one another and at differing adaptive rates.[214]

How do these theoretical concepts square with the teachings of Holy Scripture? According to Genesis, God created first humanity (Adam and Eve) in a particular place (Eden), thereby implying a monocentric

---

209. Wallbank et al., *Civilization*, 17.
210. Ibid.
211. Wood, *Human Evolution*, 86–87.
212. Ibid.
213. Ibid., 86.
214. Wallbank, Taylor, et al., *Civilization*, 17.

beginning. Later, the Adamic lineage became bottlenecked by the Noahic Flood. Only the family of Noah survived the catastrophic deluge. Therefore, the scriptural postulation is that all of current humanity derived somehow from Noahic origins (cf. Gen 9:18–19, 32). This implies a monocentric *re-beginning*. As the Genesis text affirms: "These were the sons of Noah; and from these the whole earth was peopled" (9:19) and "from these the nations spread abroad on the earth after the flood" (9:32b). It seems that the problem was that Noah's descendants, of their own volition, did not desire to spread out far enough. According to the Scriptures, humanity chose to gather and live in one narrow centralized area.

Genesis 11:1–9 presents the details of this significant post-Flood reality:

> Now the whole earth had one language and few words. And as men migrated from the east, they found a plain in the land of Shinar and settled there. And they said to one another, "Come, let us make bricks, and burn them thoroughly." And they had brick for stone, and bitumen for mortar. Then they said, "Come, let us build ourselves a city, and a tower with its top in the heavens, and let us make a name for ourselves, lest we be scattered abroad upon the face of the whole earth." And the Lord came down to see the city and the tower, which the sons of men had built. And the Lord said, "Behold, they are one people, and they have all one language; and this is only the beginning of what they will do; and nothing that they propose to do will now be impossible for them. Come, let us go down, and there confuse their language, that they might not understand one another's speech." So the Lord scattered them abroad from there over the face of all the earth, and they left off building the city. Therefore its name was called Babel, because there the Lord confused the language of all the earth; and from there the Lord scattered them abroad over the face of all the earth.

Due to another wave of human rebellion against God (this time, specifically, a resistance to the renewed divine dominion mandate given to Noah for humanity to fill the whole Earth and subdue it), God took action to force the hand of people into compliance. He "confused their [common] language" which greatly hindered their ability to communicate and work together as one efficient and effective force. Thus, the responsive result was the division of humanity into groupings (at least) along the lines of common language and the eventual dispersal of the groups into variant geographic regions: "So the Lord scattered them abroad from there over the face of all the earth" (Gen 11:9).

Again, the biblical revelation fits much more closely with the monocentric school. The current findings of science show that the oldest remains of *Homo erectus* have been found in Africa (Lake Turkana area of Kenya-Ethiopia) and are dated at 1.9 Ma. Outside of Africa, as mentioned, the oldest remains are from the Caucasus (1.8 Ma), Israel (1.5 Ma), and the Far East (1.7 Ma). There are four important points to note. First, these dates are general dates. Second, these dates are only separated by 400 thousand years (which is not all that much in relation to deep time). Third, three of the locations are in somewhat close proximity from a general geographic perspective. Fourth, all of the locations and dates could be viewed as the result of a migrational fan-out pattern. Therefore, in consideration of these points, it seems quite plausible that a monocentric origination[215] could have occurred in the Near East (i.e., from Shinar = Tigris-Euphrates River Valley based on Gen 11:2) followed by various migrational stages in multiple directions. This would be very compliant with the Babelian text of Genesis 11.

The Origin of Races

It seems very logical that the origin of "races" has a direct connection to the Babelian dispersion. The biblical reality is that there is actually no multiplicity of races. There is only one human race that comes from one human descendant by the direct act of God. In Acts 17:24–28, Paul made this statement to the Greek philosophers in Athens:

> The God who made the world and everything in it, being Lord of heaven and earth, does not live in shrines made by man, nor is he served by human hands, as though he needed anything, since he himself gives to all men life and breath and everything. And he made from one every nation of men to live on all the face of the earth, having determined allotted periods and the boundaries of

---

215. Keep in mind that we are here making reference to post-flood humanity and not to Edenic/antediluvian humanity (which, by the way, we also believe originated monocentrically). It is quite presumable that antediluvian humanity lived worldwide. This belief, of course, would require a direct refutation of the local/regional flood model (of which Ross affirms).

their habitation, that they should seek God, in the hope that they might feel after him and find him. Yet he is not far from each one of us, for "In him we live and move and have our being"; as even some of your poets have said, "For we are indeed his offspring."

Chittick asserts that the best term to describe the variations of the human race is "people groups."[216] He says further, "Physical features which are usually associated with race, such as round or slanted eyes, hair texture, and skin color are not mentioned in the Bible as race."[217] Of course, this brings to mind the important question: what caused the differentiation of physical features?

While evolutionary anthropology asserts that the random forces of natural selection are the process of differentiation, there are other postulations as well. First, there is the accentuation of certain features due to geographic and genetic isolation following Babel. Chittick explains:

> As the size of the population grew in a geographically (and genetically) isolated group, certain features were accentuated and became predominant through intermarriage within the group. Individuals who dispersed out from the Tower of Babel possessed a wide variety of genetic traits. However, no individual (or even family) possessed all genetic possibilities. After migrating out from Babel, groups settled in different areas and were somewhat geographically isolated. There was for each group a limited gene pool and intermarrying accentuated the traits of an isolated gene pool. These accentuated features are now associated with race.[218]

Second, some features—such as skin color—can become predominant in a particular area due to the level of prolonged or decreased sun exposure. Heavy exposure to sunlight can result in the loss of folic acid, which results in decreased fertility. People with darker skin have a degree of protective pigmentation against the harmful effects of the sun. People with lighter skin have much less of this protection. Therefore, in areas with heavy sun exposure, those with less protection are more susceptible to a lowered fertility rate. Over an extended period of time (i.e., multiple generations), the result would be a larger number of people with darker skin pigmentation living in that region.[219]

216. Chittick, *Puzzle of Ancient Man*, 169.
217. Ibid.
218. Ibid., 174.
219. Ibid., 172–73.

A similar yet reverse result would occur in colder areas with decreased sun exposure. Decreased exposure to sun can result in the loss of vitamin D, which can lead to rickets and other problems, including decreased fertility. People with lighter skin, even when dressed heavily to protect against the cold, have enough skin exposure to produce the vitamin D necessary to remain healthy. On the other hand, people with darker skin, having both the darker sun-filtering pigmentation along with the heavy clothing, would have a greater propensity toward a loss of vitamin D and resultant health problems, including lessened fertility. Therefore, over an extended period of time (i.e., multiple generations), the result would be a larger number of people with lighter skin living in that particular region.[220]

Third, Ross presents an important thought that keeps things in divine perspective. In the origin of the various human people groups, despite the discoveries of science, there do seem to be exceptions (e.g., dark-skinned Eskimos in the polar north and light-skinned Greeks and Egyptians in the Mediterranean) that highlight an element of mystery. Ross says:

> How did the human species develop such distinct skin colors and other more subtle differences in the relatively brief time from the days of Noah to the days of Moses? . . . At the risk of adopting what may seem to be a "God-of-the-gaps" approach, I can suggest an alternative explanation . . . Given that Genesis 11 so explicitly describes God's personal intervention in breaking up destructive unity and in motivating people to spread throughout Earth's habitable land masses, God may have done more than diversify language at that time. He possibly may have introduced also some external changes—those we recognize as racial distinctives—to facilitate the peoples' separation.[221]

He explains further:

> The two types of change [diversified language and physical distinctives] would seem to compliment each other. Just as geographical barriers (mountains, rivers, straits, swamps, and so on) and distinct languages helped move and keep the nations apart, they might be even more effective with the addition of superficial but noticeable differences of skin, hair, eyes, and so forth. Time has proven that geographical barriers by themselves do not guarantee separation, nor do distinct languages by themselves, nor do racial distinctives by themselves, but the three together erect a barely

---

220. Ibid., 173–74.
221. Ross, *Genesis Question*, 181–82.

adequate fence—at least they have until the late twentieth century. That this was God's intent seems indicated in Genesis 10:5, 20, 31 where the world's peoples are differentiated according to "their clans within their nations, each with its own language."[222]

Ross makes an important point. His postulation (as he fears) should not be disregarded as a weak "God-of-the gaps" explanation. If God caused the confusion of languages, as the Bible asserts, then Ross's postulation is no different in principle and thus quite plausible. Divine causation of the physical distinctives among people groups is a matter of primary causation (just as is the divine causation of the confusion of languages). We certainly agree with Ross that the two do seem to compliment one another. Whether his suggested explanation is true, or whether the superficial physical differentiation of various people groups is actually just the workings of genetic variability, still remains to be determined. Science may or may not eventually confirm the implicit details (e.g., genetic isolation, issues of sunlight, combination of effects, etc.) as to how human "racial" diversity happened, but regardless of such findings, that is a mere matter of secondary causation. Therefore, there is not necessarily any real conflict between supernatural involvement in the event and some sort of natural description of the event. According to Ross:

> Questions about how God introduced these changes, as well as questions about how he "confused" the languages, we cannot answer from the biblical data. Clearly, the *how* holds less importance for us than the *why*. In time, scientific research may provide at least partial answers.[223]

The *why* is indeed significant. As Ross says, "It's worth underlining that God's desire to separate the peoples of the world was due to His desires to: [1] restrain the evil that can come from a political monopoly; and [2]

---

222. Ibid., 182.

223. Ibid. Ross does provide the following scientific speculation about the *how*: "Genetic research shows the possibilities of hybridization and breed development through selective pairing. Highly selective pairing among humans might have given some impetus to the development of racial diversity. On the other hand, God might have intervened, as he seems to have done in changing telomerase activity, by miraculously introducing something new, in this case new genetic material that would generate racial distinctives. God may have used a combination of these two methods or another entirely different means, but the changes happened in a way and in a time frame that science measures as impossible by natural means alone (182)." Note that Ross's postulation fits Chittick's conception of the genetic and geographic isolation of people groups following Babel.

encourage people to 'fill the earth' and 'subdue it.'"²²⁴ It is important to understand that the divine dominion mandate is still in full force and effect today.

The Neanderthal Question

A lingering question that has long perplexed researchers is the matter of just how the Neanderthal (or Neandertal) fits into the human equation. Often called "cave men," and stereotypically pictured as big, dumb oafs, Neanderthals represent an enigma to researchers. Just who really was the Neanderthal—ape, human, some form of transitional primate hybrid, or something else entirely? Common evolutionary anthropology has typically considered Neanderthals to be either a very recent link (c. 150–30 ka) on the human evolutionary chain or an apish side branch of the evolutionary tree (diverging c. 690–550 ka) that eventually became a dead end and just disappeared (c. 30 ka).²²⁵

In 1997, genetic research caused scientists to rethink the significance of the Neanderthal. The mitochondrial DNA from several specimens, including the very first one discovered in 1856 in the Neander Valley of Germany, was tested and compared with the mtDNA of modern humans from various continents of the world. Results from the testing seemed to show that Neanderthals did not have recent linkage with humans and therefore could not have been the ancestral relatives of modern Europeans (or any other modern people group) as has often been posited.²²⁶ Mainstream science seemed to be convinced that the mystery was at long last solved. Ross has certainly taken such a position:

> Though the [prior] textbook view considers Neanderthals to be part of humanity's lineage, the preponderance of scientific evidence strongly positions them off to the side—a separate branch in any evolutionary framework . . . Instead of being rehabilitated as modern humans, the Neanderthals have been rendered extinct hominids with no bearing on humanity's origin . . . This biblical record of origins states that only one creature was made in God's

---

224. Ibid., 182.
225. Ross, *Who Was Adam?*, 179–80.
226. Serre et al., "No Evidence of mtDNA." For a detailed commentary on the implications of the tests, see also Foley, "Fossil Hominids: mitochondrial DNA."

image. That distinction belongs solely to human beings (*Homo sapiens sapiens*).[227]

In effect, Ross was asserting that Neanderthals (based on mtDNA criteria) should not be considered to be true humanity, but merely an extinct primate. However, is this assertion truly reasonable and most plausible?

Despite this alleged "preponderance of scientific evidence," some renowned paleoanthropologists, most notably Erik Trinkaus,[228] still continued to believe—on anatomical grounds—that Neanderthals had some form of direct linkage to *Homo sapiens*. Back in 1978, Trinkaus had stated: "Detailed comparisons of Neanderthal skeletal remains with those of modern humans have shown that there is nothing in Neanderthal anatomy that conclusively indicates locomotor, manipulative, intellectual or linguistic abilities inferior to those of modern humans."[229] Despite the results of the 1997 report, Trinkaus continued to assert that the genetic research would eventually confirm what he believed was the anatomical and biological "sameness" of Neanderthals and modern humans. It has long since been the contention of Trinkaus that the Neanderthal did not become extinct per se, but rather were absorbed over time through interbreeding with the Cro-Magnon masses (early modern humans—*Homo sapiens*) as they migrated from Africa into Eurasia.[230] According to Trinkaus, "When you look at all of the well-dated and diagnostic early European fossils, there is a persistent presence of anatomical features that were present among the Neandertals but absent from the earlier African modern humans. Early modern Europeans reflect both their predominant African early modern ancestry and a substantial degree of admixture between those early modern humans and the indigenous Neandertals."[231] This admixture view holds further that the behavioral difference between Neanderthals and Cro-Magnon was quite insignificant and that they understood one another to be social equals.[232] If Trinkaus is correct (and we believe that he is), it means that both groups understood one another to be human.

---

227. Ross, *Who Was Adam?*, 191.

228. Erik Trinkaus is Research Professor of Physical Anthropology at Washington University in St. Louis. He is considered to be one of the world's foremost authorities on Neanderthal biology.

229. Trinkaus, "Hard Times among the Neanderthals," 58.

230. Schoenherr, "Studies affirm relationship."

231. Ibid. Here Schoenherr quotes from his interview with Trinkaus.

232. Ibid.

The Correlation Model

Several fossils have recently been discovered that potentially provide evidence of Neanderthal/Cro-Magnon interbreeding. In 1998, the Lagar Velho child was located in an Iberian cave (in Portugal). This skeleton seems to display both Neanderthal and *Homo sapiens* anatomical characteristics.[233] Interestingly, Iberia is the last known stronghold of pure Neanderthal society.[234] Additionally, several other skeletal remains, known as the Oase fossils, were discovered incrementally in Romania between 2002 and 2005. Similarly to the Lagar Velho skeleton, these remains also show an apparent anatomical admixture of both Neanderthal and *Homo sapiens*.[235] These are not isolated cases. As Neil Schoenherr states, "From . . . other late Neanderthal sites, and recent discoveries of the earliest modern humans across Europe, a complex picture is emerging of shifting contact between behaviorally similar, if culturally and biologically different, human populations. Researchers are coming to see them all more as people, flexibly making a living through the changing human and natural landscapes of the Late Pleistocene."[236]

The strongest impetus in favor of this view comes now from the recent (2010) complete decoding of the Neanderthal genome. Richard E. Green and an international team of scientists were able to successfully sequence *the entire Neanderthal nuclear genome* (as compared with the 1997 study which was limited to just the mitochondrial DNA). The sequencing was then compared with five present-day humans from different parts of the world. The result of the study was the discovery that one to four percent of the non-African human genome are essentially from Neanderthal derivation.[237] This affirms that Trinkaus was correct all along. Green speaks to the implications:

> A striking observation is that Neandertals are as closely related to a Chinese and Papuan individual as to a French individual, even though morphologically recognizable Neandertals exist only in the fossil record of Europe and western Asia. Thus, the gene flow between Neandertals and modern humans that we detect most likely occurred before the divergence of Europeans, East Asians,

---

233. Lubenow, "Lagar Velho 1 child skeleton," 6–8.

234. Hall, "Last of the Neanderthals," 55–59. See also, Tattersal and Schwartz, "Hominids and hybrids," 7117.

235. Fitzpatrick, "A jaw-some discovery."

236. Schoenherr, "Late Neanderthals and modern human contact in southeastern Iberia."

237. Green et al., "Draft Sequence of the Neandertal Genome," 710, 721.

and Papuans. This may be explained by mixing of early modern humans ancestral to present-day non-Africans with Neandertals in the Middle East before their expansion into Eurasia.[238]

While anthropology seeks to explain humanity in terms of evolutionary divergence, we suggest that the Neanderthal relationship to Europeans, East Asians, Papuans, etc. is precisely what should be expected following the (monocentric) Babelian dispersion of Genesis 11 (which is from the Middle East). If Neanderthals were of the human genus, then they should be "as closely related to a Chinese and Papuan individual as to a French individual" due to their common Middle Eastern beginnings and ultimate Eurasian geographic concentration and genetic isolation. It should also be noted that Neanderthals are not genetically unrelated to the people groups of Africa; they are merely related to a lesser degree.[239] This is compatible with the teachings of Holy Scripture. After all, based on the biblical revelation, all humans have a common Adamic origin (and all post-flood humans have a common Adamic-Noahic origin). We posit that Neanderthals were simply a people group that migrated from Babel and concentrated primarily in Eurasia (and eventually were absorbed by other people groups who later advanced into that same region). The Neanderthals, who probably never exceeded a peak population of 15,000 in Western Europe,[240] eventually "disappeared" in the midst of the massive Cro-Magnon migration from Africa. Therefore, being only a few drops in the greater sea of humanity, it is not surprising that the Neanderthal genome is only one to four percent present in the contemporary human gene pool (although Trinkaus reminds all researchers that that is merely a minimum).[241] The key reality is that the Neanderthal genome *is* measurably present, particularly in the non-African population. This implies that Neanderthals (contrary to the assertions of Ross) were not some form of advanced ape primates, but true human beings blessed with the image of God.

As to the "cave man" image, there is certainly some truth involved—but not as commonly stereotyped. After Babel, the Neanderthal people groups

---

238. Ibid., 721.

239. Ibid. Green states: "Neandertals are on average closer to individuals in Eurasia than to individuals in Africa."

240. Hall, "Last of the Neanderthals," 38.

241. Than, "Neanderthals, Humans Interbred." Here Trinkaus comments as to the presence of the Neanderthal genome in modern humanity: "One to four percent is truly a minimum. But is it ten percent? Twenty percent? I have no idea."

would have experienced at least two major challenges. First, they would have experienced *culture loss*. As they traveled away from the Middle East and further into geographic isolation, their immediate primary goal would have been survival—food, clothing, and safety—not cultural and technological preservation and development. Any matters of cultural and technological practice would have been primarily related to the perseverance of the group. Secondly, they would have experienced *harsh living conditions*. As they migrated primarily into the new frontier of Eurasia, especially into zones of expansive Ice Age glaciation, conditions would have been quite difficult. As Chittick comments:

> In this geographical area, climate and living conditions were harsh because of the advancing Ice Age at that time. In order to cope with the harsh weather, caves were utilized as shelters and living quarters. These early settlers at the edge of the ice were "cave men." May I suggest therefore that "cave man" was not primitive man, but rather, cave man was de-cultured man. They (the cave dwellers) were intellectually quite capable but had suffered cultural loss and would incorrectly be termed aboriginal or primitive by those who believe that man had evolved up from an animal.[242]

The status of the Neanderthal is a very important matter in understanding the human genus from a biblical worldview. We hold that they are one prominent example of human diversification significantly influenced by group migrational isolation.

The Big Picture of Humanity

Humanity—created late in God Day Six, yet primarily existing in God Day Seven, and dating back to *Homo erectus* beginnings (c. 1.9 Ma—see Wood and Collard)—has shown no major cladogenesis events or speciation in the *Homo erectus*-*Homo sapiens* relationship (see Wolpoff), although common anthropology asserts that there has been polytypic regional diversification (i.e., *Homo ergaster, Homo heidelbergensis, Homo neanderthalensis*) over time. This signifies a long-lasting and connective human lineage to the present day. To better understand the *Homo* genus, we suggest that a traditional approach be used (i.e., grouping the diverse human taxa, prior to modern *Homo sapiens*, together as "archaic" *Homo sapiens*—we actually

---

242. Chittick, *Puzzle of Ancient Man*, 165.

prefer "early" *Homo sapiens*)²⁴³ with slight modification. We modify with the further suggestion that all polytypic variations of the *Homo* genus, beginning with *Homo erectus*, be classified specifically as *Homo sapiens* (i.e., *Homo sapiens erectus*, *Homo sapiens ergaster*, *Homo sapiens heidelbergensis*, *Homo sapiens neanderthalensis*, *Homo sapiens sapiens*) to show the continuity amidst any perceived technical anatomical differentiation. Though we believe this to be very compatible with the scriptural revelation, it certainly flies in the face of some evolutionary anthropologists,²⁴⁴ who believe that the best approach is to create additional differentiating taxa to clarify the morphological complexity (as opposed to unifying the taxa). Of course, the maxi-taxa approach is overtly macroevolutionary and finds itself in direct contrast to an evangelical Christian worldview. The macroevolutionary mindset is one that constantly assumes and seeks evidence of evolutionary transitionalism in the fossil record, namely, the more taxa, the more transitionalism. Therefore, it is natural with such a worldview to strongly emphasize characteristics that overtly differentiate the human fossils and propose taxonomic solutions commensurate to that end.

On the other hand, from the biblical view of Adamic origination, though there exists physical differentiation within the human genus, we contend that humanity did not macroevolve from lower taxa to higher taxa. Humanity was created directly by God and has been bearing of his image from the Adamic inception. A much better explanation is that God specially created humanity as distinctly different from all other creatures. This differentiation includes advanced intelligence, the capacity for an ordered society, and a corresponding moral-ethical responsibility.²⁴⁵ The past and

---

243. For disagreement with this approach, see Rightmire, "Human Evolution in the Middle Pleistocene," 218–27. He states: "This Middle Pleistocene record, still sparse but increasingly well dated, raises important questions. One concerns the fate of *Homo erectus* in different regions of the Old World. Another is how many distinct species should be recognized among the descendants of this ancient lineage. It is apparent that the traditional approach of lumping diverse humans together as 'archaic' *Homo sapiens* will no longer work. The picture is highly complex, and several taxa probably are needed to accommodate the fossils. Evolutionary relationships among these populations must be clarified, but pose some major problems" (216).

244. Ibid.

245. For a good discussion of the theological consequences of the macroevolution of humanity, see Young, *Creation and the Flood*, 155–58. According to Young: "Acceptance of an evolutionary view of the origin of man thus introduces a distortion to the gospel. Sin is not a transgression to the law of God, nor a violation of the original nature of man. Sin is now really nothing more than an unpleasant characteristic or action that it would be nice to eliminate from society. The sinner, when confronted with the fact of

current physical diversity of people has occurred through a guided process over time involving some form of divine intervention (e.g., possible introduction of new superficial genetic material, etc.) along with scientifically explainable genetic (e.g., selective pairing, variability, etc.) and geographic (e.g., regional isolation, etc.) factors.[246] This diversity seems to have been a particularly evident phenomenon since Babel. Yet, despite the human diversity (from early *Homo erectus* to modern *Homo sapiens*), people are still people and have always been people.

THE CONCLUSION

We have determined that a scripturally-based Old-Earth Creationism paradigm and the geologic ages as posited by modern science can indeed be found to be both compatible and veracious. To show this, we have presented a model correlating the God Days of Genesis with the geologic column. The biblical description of each God Day (Days 1–7) shows the progressive "day to day" high points of the divine creation and sustaining process. Meanwhile, we have shown that science can corroborate the biblical revelation and often provide additional detail. We have sought to clearly convey that the biblical revelation, when properly understood within the

---

his sinfulness, can with a considerable degree of legitimacy make excuses for his sin. He can to a large degree blame his behavior on his biological heritage. He can even to some extent lay the blame, according to the theistic evolutionist construction, on God. After all, has not God directed the process which has given him a certain biochemical structure together with all of its instinctual behavior patterns? How then can the sinner truly be guilty before God when sin is built into his very structure?" (157–58). Moreover, Young adds: "There also remains the further problem, in the theistic evolutionist interpretation, of the meaning of Christ's death. What is meant by the statement that Christ died for our sin? Careful reading of the Bible, and particularly the New Testament, ought to make very plain that Christ's death on the cross did not change our biochemical structure so that we no longer follow certain instinctive behavioral patterns. In fact it ought to be clear Christ died so that His people might have their sins forgiven by God, and that through His Spirit He might renew the innermost being of His people by giving to them a new heart of flesh and spirit. Adopting a theistic evolutionist approach to the origin of man places the entire structure of Christian theology and ethics in a most precarious position" (158).

246. Ibid., 155. Young comments: "Perhaps the most useful task for the Christian paleoanthropologist would be to evaluate the available fossil material carefully and cautiously and to show that the evidence for an evolution from ape to man is much more tenuous than we are sometimes led to believe. It might be learned that man has simply undergone a small degree of physical change through time and that physically he is much more variable than was hitherto thought."

bounds of certain important presuppositions, including, most notably, the divinely-ordained Anthropic Principle, does not conflict with the valid postulations of empirical science. It is our view that the addition of this correlation model to the tenets of Old-Earth Progressive Creationism greatly strengthens its overall integrity and its conformity to composite reality.

Bibliography

Adams, Robert M. "Theodicy." In *The Cambridge Dictionary of Philosophy*, edited by Robert Audi. 910–11. New York: Cambridge University Press, 2005.
Ager, Derek V. *The Nature of the Stratigraphical Record*. Chichester, England: John Wiley & Sons, 1993.
———. *The New Catastrophism*. Cambridge: Cambridge University Press, 1995.
Bajpai, Sunil, and Philip D. Gingerich. "A new Eocene archaeocete (Mammalia, Cetacea) from India and the time of origin of whales." In *Proceedings of the National Academy of Sciences* 26 (December 1998) 15464–68.
Bates, Robert, and Julia A. Jackson. *Dictionary of Geological Terms*. New York: Anchor, 1984.
Batten, Don, and Jonathan Sarfati. *15 Reasons to Take Genesis as History*. Brisbane, AU: Creation Ministries International, 2006.
Berner, Robert A. "Atmospheric oxygen over Phanerozoic time." In *Proceedings of the National Academy of Sciences of the United States of America* 96 (September 1999) 10956–57.
Bird, W. R. *The Origin of Species Revisited* (two volumes). Vol. 1. New York: Philosophical Library, 1989.
Blackman, Cyril E. "Romans." In *The Interpreter's One-Volume Commentary on the Bible*, edited by Charles M. Laymon, 784. Nashville: Abingdon, 1971.
Brown, Francis, S. R. Driver, et al. *The Brown-Driver-Briggs Hebrew and English Lexicon*. Peabody, MA: Hendrickson, 2006.
Buswell, J. Oliver, Jr. "Creation Days." In *Journal of the American Scientific Affiliation* 4 (March 1952) 10–13.
Chaffey, Tim, and Jason Lisle. *Old-Earth Creationism on Trial*. Green Forest, AR: Master, 2008.
Chen, Chin-Wu, Stephane Rondenay, et al. "Geophysical Detection of Relict Metasomatism from an Archean (~3.5 Ga) Subduction Zone." *Science* 326 (November 2009) 1089–91.
Chittick, Donald E. *The Puzzle of Ancient Man*. Newberg, OR: Creation Compass, 2006.
Cifelli, Richard L. "Early Mammalian Radiation." *Journal of Paleontology* 75 (November 2001) 1214–26.
Clark, Harold W. *Genesis and Science*. Nashville: Southern, 1967.
Collins, C. John. *Genesis 1–4*. Phillipsburg, NJ: P&R, 2006.
———. *Science & Faith: Friends or Foes?* Wheaton, IL: Crossway, 2003.
Collins, Francis S. *The Language of God*. New York: Free, 2006.

BIBLIOGRAPHY

Cragg, Gerald R. "Romans." Exegesis Section. In *The Interpreter's Bible* (twelve volumes), edited by George Arthur Buttrick, Vol. 9, 521. Nashville: Abingdon, 1954.

Craig, William Lane. *Time and Eternity.* Wheaton, IL: Crossway, 2001.

———. "Time, Eternity, And Eschatology." In *The Oxford Handbook of Eschatology,* edited by J. Walls. 596–613. Oxford, England: Oxford University Press, 2008.

———. "The Ultimate Question of Origins: God and the Beginning of the Universe." Accessed March 10, 2008. https://www.reasonablefaith.org/writings/scholarly-writings/the-existence-of-god/the-ultimate-question-of-origins-god-and-the-beginning-of-the-universe/.

Cremo, Michael A., and Richard L. Thompson. *Forbidden Archaeology.* Los Angeles: Bhaktivedanta, 1998.

Cuoghi, Diego. "The Mysteries of the Piri Reis Map." Accessed May 22, 2010. http://xoomer.virgilio.it/dicuoghi/Piri_Reis/PiriReis_eng.htm.

Cvancarra, Alan M. *A Field Manual for the Amateur Geologist.* San Francisco: Jossey-Bass, 1995.

Dehaan, Robert F., and John L. Wiester. "The Cambrian Explosion." In *Signs of Intelligence,* edited by William A. Dembski and James M. Kushiner. Grand Rapids: Brazos, 2005.

Dembski, William A. *The End of Christianity* Nashville: B&H, 2009.

DiSalle, Robert. "Space-time." In *The Cambridge Dictionary of Philosophy,* edited by Robert Audi, 867. New York: Cambridge University Press, 1999.

Doukhan, Jacques B. "The Genesis Creation Story: Text, Issues, and Truth." *Origins* 55 (2004) 12–33.

Dudley, Robert. "Atmospheric Oxygen, Giant Paleozoic Insects and the Evolution of Aerial Locomotor Performance." In *The Journal of Experimental Biology* 201 (1998) 1043–50.

Durant, Will, and Ariel Durant. *The Lessons of History.* New York: Simon and Schuster, 1968.

———. *Our Oriental Heritage.* The Story of Civilization, vol. 1, (eleven volumes). New York: Simon And Schuster, 1954.

Dutch, Steven. "The Piri Reis Map." Accessed May 22, 2010. http://www.uwgb.edu/dutchs/pseudsc/pirireis.htm.

Earman, John and Richard M. Gale. "Time." In *The Cambridge Dictionary of Philosophy,* edited by Robert Audi. 920. New York: Cambridge University Press, 1999.

Eerdman, Cordelia. "Fossil Sequence in Clearly Superimposed Rock Strata." *Journal of the American Scientific Affiliation* 2 (1950) 13–17.

———. "Stratigraphy and Paleontology." In *Journal of the American Scientific Affiliation* 5 (March 1953) 3–6.

Emberger, Gary. "Theological and Scientific Explanations for the Origin and Purpose of Natural Evil." *Perspectives on Science and Christian Faith* 46 (September 1994) 150–58.

Erickson, Millard J. *Christian Theology.* Baker, 1993.

Finlay, Graeme. "*Homo divinus*: The ape that bears God's Image." *Science and Christian Belief* 15 (April 2003) 17–40.

Fischer, Robert B. *God Did It, But How?* Ipswich, MA: ASA, 2005.

Fitzpatrick, Tony. "A jaw-some discovery: Earliest modern human fossils in Europe found in bear cave." *The Source,* January 1, 2002. http://news.wustl.edu/news/Pages/2370.aspx.

Bibliography

Foley, Jim. "Fossil Hominids: mitochondrial DNA" Accessed June 30, 2010. http://www.talkorigins.org/faqs/homs/mtDNA.html.

Fretwell, P., H. D. Pritchard, et al. "Bedmap2: improved ice bed, surface and thickness datasets for Antarctica." *The Cryosphere* 7 (28 February 2013) 375-93.

Froade, Carl, Jr. *Geology By Design*. Green Forest, AR: Master, 2007.

———. "Radiometric Cherry-Picking." *Creation Matters* 15 (November/December 2010) 1–3.

Gibbard, Philip L., Martin J. Head, and Michael J. C. Walker. "Formal Ratification of the Quaternary System/Period and the Pleistocene Series/Epoch with a base at 2.58 Ma." *Journal of Quaternary Science* 25 (September 2009) 96–102.

Gibbons, Ann. "In Search of the First Hominids." *Science* 295 (February 15, 2002) 1214–19.

Gibson, L. James. "Polyphyly and the Cambrian Explosion." *Origins* 52 (2001) 3–6.

Godfrey, Stephen J. and Christopher R. Smith. *Paradigms on Pilgrimage*. Toronto: Clements, 2005.

Goodwin, Anna, Jon Wyles, et al. "The Permo-Triassic Mass Extinction." University of Bristol Department of Earth Sciences—Paleobiology and Biodiversity Research Group (2001). http://archive.fo/hOY0.

Green, Richard E., Johannes Krause, et al. "A Draft Sequence of the Neandertal Genome." *Science* 328 (May 7, 2010) 710–22.

Green, William Henry. "Primeval Chronology." *Bibliotheca Sacra* (April 1890) 285–303. http://www.girs.com/library/theology/syllabus/creation_green.html.

Hagee, John. *The Revelation of Truth*. Nashville: Thomas Nelson, 2000.

Hall, Stephen S. "Last of the Neanderthals." *National Geographic* 214 (October 2008) 55–59.

Hapgood, Charles H. *Maps of the Ancient Sea Kings*. Kempton, IL: Adventures Unlimited, 1996.

Harrison, Everett F. "Romans." In *The Expositor's Bible Commentary* (twelve volumes), edited by Frank E. Gaebelein. Vol. 10, 94. Grand Rapids: Zondervan, 1976.

Hauser, Marc, Rossana Martini, et al. "The break-up of East Gondwana along the northeast coast of Oman: evidence from the Batain basin." *Geological Magazine* 139 (March 2002) 145–57.

Hawking, Stephen W. *A Brief History of Time*. New York: Bantam, 1988.

Hayward, James L., and Donald E. Casebolt. "The Genealogies of Genesis 5 and 11: A Statistical Study." *Origins* 9 (1982) 75–81.

———. "Reactions." *Origins* 10 (1983) 5–8.

Heckman, D. S., D. M. Geiser, et al. "Molecular evidence for the early colonization of land by fungi & plants." *Science* 293 (2001) 1129–33.

Hefferan, Kevin. "Archean Life." University of Wisconsin-Stevens Point, Accessed March 24, 2009. https://www4.uwsp.edu/geo/faculty/hefferan/geol106/class3/ARCHEAN%20LIFE.htm.

Hick, John. *Evil and the God of Love*. New York: Harper & Row, 1966.

Hinshaw, Gary F., J. L. Weiland, et al. "Five-Year Wilkinson Microwave Anisotrophy Probe (WMAP) Observations: Data Processing, Sky Maps, & Basic Results." *Astrophysical Journal Supplement Series* 180 (2009 February) 225–45.

Hyers, Conrad. *The Meaning of Creation: Genesis and Modern Science*. Atlanta: John Knox, 1984.

Bibliography

Ingolfsson, Olafur. "Quaternary glacial and climate history of Antarctica." In *Extent and Chronology of Glaciation, Vol. 3: South America, Asia, Africa, Australia, and Antarctica*, edited by Jurgen Ehlers and Philip L. Gibbard, 3–43. Amsterdam: Elsevier Science, 2004.

Johnson, Gaines R. *The Bible, Genesis & Geology*. Charleston, SC: CreateSpace, 2010.

Kazlev, M. Alan. "The Archean Eon." *Palaeos: Life Through Deep Time*. Accessed March 24, 2009. http://www.palaeos.com/Archean/Archean.htm.

———. "The Carboniferous." *Palaeos: Life Through Deep Time*. Accessed March 24, 2009. http://palaeos.com/paleozoic/carboniferous/carboniferous.html

———. "Gondwana." *Palaeos: Life Through Deep Time*. Accessed March 25, 2009. http://palaeos.com/earth/paleogeography/gondwana.html.

Kazlev, M. Alan, and Franco Maria Boschetto. "The Hadean Eon." *Palaeos: Life Through Deep Time*. Accessed March 24, 2009. http://palaeos.com/hadean/hadean.html.

Klotz, John W. "The Philosophy of Science in Relation to Concepts of Creation vs. The Evolution Theory." In *Why Not Creation?*, edited by Walter E. Lammerts, 5–23. Grand Rapids: Baker, 1976.

Kohler, Kaufman and Emil G. Hirsch. "Creation." In *The Jewish Encyclopedia*, edited by Isidore Singer, 4:336–40. New York: Funk and Wagnalls, 1909.

Lambert, David. *The Field Guide to Geology*. New York: Checkmark, 2007.

Lammerts, Walter E. "Introduction." In *Why Not Creation?*, edited by Walter E. Lammerts, 1–4. Grand Rapids: Baker, 1970.

Lane, Nick. "First Breath: Earth's billion-year struggle for oxygen." *New Scientist* (February 6, 2010) 36–39.

Leakey, Richard E. *The Making of Mankind*. New York: Dutton, 1981.

Leakey, Richard E., and Roger Lewin. *Origins Reconsidered*. New York: Anchor, 1993.

Lerner, K. Lee, and Brenda Wilmoth Lerner. "Supercontinents." In *World of Earth Science*, Vol. 2. Farmington Hills, MI: Gale Cengage, 2003.

Lee, Michael S. Y., Andrea Cau, et al. "Morphological Clocks in Paleontology, and a Mid-Cretaceous Origin of Crown Aves." *Systematic Biology* 63 (2014) 442–49.

Leupold, H. C. *Exposition of Genesis I*. Grand Rapids: Baker, 1949.

Lindenfors, Patrik, and Birgitte S. Tullberg. "Phylogenetic Analyses of Primate Size Evolution: The Consequences of Sexual Selection." *Biological Journal of the Linnean Society* 64 (1998) 413–47.

Lubenow, Marvin L. "Lagar Velho 1 child skeleton: a Neanderthal/modern human hybrid." *CEN Technical Journal* 14 (2000) 6–8.

MacArthur, John. *The Battle for the Beginning*. Nashville: Thomas Nelson, 2001.

McGowan, Brian, Bill Berggren, et al. "Neogene and Quaternary coexisting in the geological time scale: The inclusive compromise." *Earth-Science Reviews* 96 (June 2009) 249–62.

McPhee, John. *Basin and Range*. New York: Farrar, Straus, and Giroux, 1981.

McTaggert, J. M. E. *The Nature of Existence*. Vol. 1. Cambridge: Cambridge University Press, 1927.

———. "The Unreality of Time." In *The Philosophy of Time*, edited by Robin Le Poidevin and Murray MacBeath. 23–34. Oxford: Oxford University Press, 1993.

Miller, Keith B. "And God Saw That It Was Good: Death and Pain in the Created Order." *Perspectives on Science and Christian Faith* 63 (June 2011) 85–93.

Miller, Kenneth R. *Finding Darwin's God*. New York: Harper Perennial, 1999.

Bibliography

Mohler, R. Albert, Jr. "Christianity and Evolution—Seeing the Problem." February 16, 2009. http://www.albertmohler.com/2009/02/16/christianity-and-evolution-seeing-the-problem/.
Monastersky, Richard. "Ancient animals got a rise out of oxygen—Carboniferous period invertebrates." *Science News* (May 13, 1995) 1–2.
Moreland, J. P. "Introduction." In *The Creation Hypothesis*, edited by J. P. Moreland, 11. Downer's Grove, IL: InterVarsity, 1994.
Morris, Henry M. and Gary E. Parker. *What Is Creation Science?* Green Forest, AR: Master, 1987.
Morris, Henry III. "Being Like Him." Institute for Creation Research. December 2009. http://www.icr.org/articles/print/5033/.
———. "The Issues of Death." In *Acts & Facts* 38 (November 2009) 22.
Morton, Glenn R. "The Geologic Column and its implications for the Flood." *Talk Origins Archive* (February 17, 2001) 1–15.
Mounce, William D. *Mounce's Complete Expository Dictionary of Old & New Testament Words*, edited by William D. Mounce. Grand Rapids: Zondervan, 2006.
Murck, Barbara. *Geology*. New York: John Wiley & Sons, 2001.
Murphy, Dennis C. "Archaeopteris: The First Modern Tree." *Devonian Times*. Updated July 9, 2005. http://www.devoniantimes.org/who/pages/archaeopteris.html.
———. "Early Seed Plants: The Start of Something Big." *Devonian Times*. Updated July 9, 2005. http://www.devoniantimes.org/who/pages/lyginopterids.html.
Murphy, J. B., and R. D. Nance. "Earth Sciences: On the Assembly of Supercontinents." *American Scientist* 92 (2004) 324.
Nance, R. D., T. R. Worsley, et al. "The Supercontinent Cycle." *Scientific American* 259 (July 1, 1988) 72–79.
Newman, Robert C., John A. Bloom, et al. "The Status of Evolution as a Scientific Theory." (November 3, 2001). http://www.arn.org/docs/newman/rn_statusofevolution.htm.
Newman, Robert C., and Herman J. Eckelmann, Jr. *Genesis One and the Origin of the Earth*. Hatfield, PA: Interdisciplinary Biblical Research Institute, 1977.
Neyman, Greg. "Death Before the Fall of Man." *Old Earth Ministries*. First published January 28, 2003. http://www.oldearth.org/print/death.pdf.
Nield, Ted. *Supercontinent*. Cambridge, MA: Harvard University Press, 2007.
Noorbergen, Rene. *Secrets of the Lost Races*. Brushton, NY: Teach, 2006.
———. *Treasures of the Lost Races*. Brushton, NY: Teach, 2004.
Numbers, Ronald L. *The Creationists*. Cambridge, MA: Harvard University Press, 2006.
Oard, Michael J. "A Post-Flood Ice-Age Model Can Account For Quaternary Features." *Origins* 17 (1990) 8–26.
Ogg, James G., Gabi Ogg, et al. *The Concise Geologic Time Scale*. Cambridge: Cambridge University Press, 2008.
O'Neil, Dennis. "Early Modern Homo Sapiens." Palomar College, Department of Anthropology (March 2010) http://anthro.palomar.edu/homo2/mod_homo_4.htm.
Phillips, Perry G. "Did Animals Die before the Fall?" *Perspectives on Science and Christian Faith* 58 (June 2006) 146–47.
Pierce, Larry and Ken Ham. "Are There Gaps in the Genesis Genealogies?" In *The New Answers Book 2*. Green Forest, AR: Master, 2010.
Plantinga, Alvin. "Methodological Naturalism?" *Perspectives on Science and Christian Faith* 49 (September 1997) 143–54.

Bibliography

Premovic, Pavle I. "The late Paleozoic oxygen pulse and accumulations of petroleum source rocks and coal." In *Journal of the Serbian Chemical Society* 71 (2006) 143–47.

Ramm, Bernard. *The Christian View of Science and Scripture*. Grand Rapids: Eerdmans, 1978.

Rightmire, G. Philip. "Human Evolution in the Middle Pleistocene: The Role of *Homo Heidelbergensis*." In *Evolutionary Anthropology* 6 (1998) 218–27.

Ritland, Richard. "Historical Development of the Current Understanding of the Geologic Column: Part I." *Origins* 8 (1981) 59–76.

———. "Historical Development of the Current Understanding of the Geologic Column: Part II." *Origins* 9 (1982) 28–50.

Robertson Group. *Stratigraphic Database of Major Sedimentary Basins of the World*. Llandudno Gwynedd, UK: The Robertson Group, 1989.

Robinson, Theodore H. "Genesis." In *The Abingdon Bible Commentary*, edited by Frederick Carl Eiselen, Edwin Lewis, et al., 220. Nashville: Abingdon, 1929.

Ross, Hugh. *Creation as Science*. Colorado Springs, CO: NavPress, 2006.

———. *The Genesis Question*. Colorado Springs, CO: NavPress, 2001.

———. *A Matter of Days*. Colorado Springs, CO: NavPress, 2004.

Roth, Ariel A. "Does Evolution Qualify as a Scientific Principle?" *Origins* 4 (1977) 4–10.

Rubey, William W. "The Geologic History of Sea Water—An Attempt to State the Problem." *Geological Society of America Bulletin* 62 (1951) 1111–48.

Rudwick, Martin J. S. *Bursting the Limits of Time*. Chicago: The University of Chicago Press, 2005.

———. *Worlds before Adam*. Chicago: The University of Chicago Press, 2008.

Rusbult, Craig. "Do we have evidence for an old earth-and universe?" *The American Scientific Affiliation* (2006). http://www.asa3.org/ASA/education/origins/agescience.htm.

Sailhamer, John H. "Genesis." In *The Expositor's Bible Commentary* (twelve volumes), edited by Frank E. Gaebelein, vol. 2, 24–38. Grand Rapids: Zondervan, 1996.

Saniga, Metod. "Geometry of Time and Dimensionality of Space." In *The Nature of Time: Geometry, Physics, & Perception*, edited by R. Buccheri, M. Saniga, et al., 131–43. NATO Science Series II—Dordrecht, NL: Kluwer, 2003.

Sarfati, Jonathan. *Refuting Compromise*. Green Forest, AR: Master, 2004.

Schaeffer, Francis A. *Genesis in Space and Time* Downer's Grove, IL: InterVarsity, 1972.

Schicatano, Jim. *The Theory of Creation*. San Jose, CA: Writers Club, 2001.

Schmidt, Gavin A. and Adam Frank. "The Silurian Hypothesis: Would It Be Possible to Detect an Industrial Civilization in the Geological Record?" In *International Journal of Astrobiology* (2018) 1–9.

Schoenherr, Neil. "Late Neanderthals and modern human contact in southeastern Iberia." *The Source*, December 10, 2008. Washington University in St. Louis. http://news-info.wustl.edu/tips/page/normal/13145.html.

———. "Studies affirm relationship between early humans, Neanderthals." *The Source*, June 14, 2007. Washington University in St. Louis. http://news-info.wustl.edu/tips/page/normal/9520.html.

Schrader, Stephen R. "Genesis." In *The Parallel Bible Commentary*, edited by Edward E. Hindson and Woodrow Michael Kroll. Nashville: Thomas Nelson, 1994.

Scott, Charles Anderson. "Romans." In *The Abingdon Bible Commentary*, edited by Frederick Carl Eiselen, Edwin Lewis, et al., 1154. Nashville: Abingdon, 1929.

BIBLIOGRAPHY

Scott, Eugenie C. *Evolution Vs. Creationism: An Introduction*. University of California Press, 2005.

Seely, Paul. H. "Adam and Anthropology: A Proposed Solution." In *Journal of the American Scientific Affiliation* 22 (September 1970) 88–90.

Serre, David, Andre Langaney, et al. "No Evidence of mtDNA Contribution to Early Modern Humans." In *PloS Biology Journal* 2 (March 16, 2004) 57.

Sewell, Curt. "Uniformitarianism and the Geologic Column." November 8, 1999. Revolution Against Evolution. http://www.rae.org/bits12.htm.

Shen, Bing, Lin Dong, et al. "The Avalon Explosion: Evolution of Ediacara Morphospace." *Science* 319 (January 2008) 81–84.

Shipman, Pat. "Hunting the First Hominid." In *American Scientist* 90 (Jan/Feb 2002) 25.

Simpson, Cuthbert A. "Genesis" Exposition Section. In *The Interpreter's Bible* (twelve volumes), edited by George Arthur Buttrick, Vol. 1, 469. Nashville: Abingdon-Cokesbury, 1952.

Snoke, David. *A Biblical Case for an Old Earth*. Grand Rapids: Baker, 2006.

Southworth, C. Scott. Reston, VA: U.S. Geologic Survey. Personal email. February 17, 2011.

Speer, Brian R. "Introduction to the Archean." University of California-Berkeley Museum of Paleontology. Updated July 7, 2011. http://www.ucmp.berkeley.edu/precambrian/archean.html.

Stearley, Ralph. "Assessing Evidences for the Evolution of a Human Cognitive Platform for 'Soulish Behaviors.'" In *Perspectives on Science and Christian Faith* 61 (September 2009) 152–74.

Stearn, Colin and Robert Carroll. *Paleontology*. New York: Wiley & Sons, 1989.

Stiebing, Jr., William H. *Ancient Astronauts, Cosmic Collisions and Other Popular Theories about Man's Past*. Amherst, NY: Prometheus, 1984.

Steur, Hans. "Fossil Plants." Hans' Paleobotany Pages. Updated May 10, 2017. http://www.xs4all.nl/~steurh/home.html.

———. "How Plants Conquered the Land." Hans' Paleobotany Pages. Updated May 10, 2017. http://www.xs4all.nl/~steurh/eng/old1.html.

Stoner, Don. *A New Look at an Old Earth*. Eugene, OR: Harvest House, 1997.

Sutera, Raymond. "The Origin of Whales and the Power of Independent Evidence." In *Reports of the National Center for Science Education* 20 (Sept/Oct 2000) 33–41.

Swinburne, Richard. *The Existence of God*. Oxford: Clarendon, 1991.

Tamura, Nobu. "Xiaotingia zhengi or is Archaeopteryx still a bird?" *Paleoexhibit* (August 2011) 1–3. http://paleoexhibit.blogspot.com/2011/08/xiaotingia-zhengi-or-is-archaeopteryx.html.

Tattersal, Ian and Jeffrey H. Schwartz. "Hominids and hybrids: The place of Neanderthals in human evolution." *Proceedings of the National Academy of Sciences* 96 (June 1999) 7117.

Tegmark, Max. "On the Dimensionality of Space-Time." *Classical and Quantum Gravity* 14 (April 1997) L69-L75.

Than, Ker. "Neanderthals, Humans Interbred—First Solid DNA Evidence." *National Geographic News*, May 6, 2010. http://news.nationalgeographic.com.

Thewissen, J. G. M. and S. Bajpai. "Whale Origins as a Poster Child for Macroevolution." *BioScience* 51 (December 2001) 1037–49.

Trinkaus, Eric. "Hard Times among the Neanderthals." *Natural History* 87 (December 1978) 58.

BIBLIOGRAPHY

Turney, Chris. *1912: The Year The World Discovered Antarctica.* Berkeley, CA: Counterpoint, 2012.

Tyler, David J. and Harold G. Coffin. "Accept the Column, Reject the Chronology." In *The Geologic Column,* edited by John K. Reed and Michael J. Oard, 53–68. Chino Valley, AZ: Creation Research Society, 2006.

Vardiman, Larry. *Ice Cores and the Age of the Earth.* El Cajon, CA: Institute for Creation Research, 2003.

Vine, W. E., Merrill Unger, et al. *Vine's Complete Expository Dictionary of Old and New Testament Words.* New York: Thomas Nelson, 1985.

Waggoner, Ben M. "Hadean time: 4.5 to 3.8 billion years ago." University of California-Berkeley Museum of Paleontology. Updated July 7, 2011. http://www.ucmp.berkeley.edu/precambrian/hadean.html.

Wallbank, T. Walker, Alastair M. Taylor, et al. *Civilization: Past & Present.* Volume 1. Glenview, IL: Scott, Foresman, 1976.

Walton, John H., Victor H. Matthews, et al. "Genesis." In *The IVP Bible Background Commentary.* OT Volume. Downer's Grove, IL: IVP Academic, 2000.

Weisburd, Stefi. "A forest grows in Antarctica—an extensive forest may have flourished about 3 million years ago." *Science News* (March 8, 1986) 40–43.

Weisz, Paul B. and Richard N. Keogh. *The Science of Biology.* New York: McGraw-Hill, 1982.

Wellman, C. H., P. L. Osterff, et al. "Fragments of the earliest land plants." *Nature* 425 (2003) 282–85.

Wesley, John. "The General Deliverance." In *The Works of John Wesley,* edited by Albert C. Outler. Vol. 1, 437–50. Nashville: Abingdon, 1985.

———. "The New Creation." In *The Works of John Wesley,* edited by Albert C. Outler, Vol. 2, 500–510. Nashville: Abingdon, 1985.

Wheeler, Gerald. "The Cruelty of Nature." *Origins* 2 (1975) 32–41.

White, A. Toby. "The Proterozoic Eon of Precambrian Time: 2500 to 542 million years ago." *Palaeos: Life Through Deep Time.* Accessed March 25, 2009. http://palaeos.com/proterozoic/proterozoic.html.

Wiens, Roger C. "Radiometric Dating: A Christian Perspective." *American Scientific Affiliation* 3 (2002) 1–33.

Willimon, William H. *Sighing for Eden* Nashville: Abingdon, 1985.

Wilson, Marvin R. *Exploring Our Hebraic Heritage.* Grand Rapids: Eerdmans, 2014.

Wilson, William. *Old Testament Word Studies.* Grand Rapids: Dregel, 1978.

Woese, Carl and J. Peter Gogarten. "When did eukaryotic cells (cells with nuclei and other internal organelles) first evolve? What do we know about how they evolved from earlier life forms?" *Scientific American* (October 21, 1999). http://www.scientifcamerican.com/article.cfm?id=when-did-eukaryotic-cells.

Wolpoff, Milford H., Alan G. Thorne, et al. "The Case for Sinking *Homo Erectus,* 100 Years of *Pithecanthropus* is Enough!" In *Courier Forschungs-Institut Senckenberg* 171 (January 1994) 341–61.

Wonderly, Daniel E. *God's Time-Records in Ancient Sediments.* Hatfield, PA: Interdisciplinary Biblical Research Institute, 1999.

Wood, Bernard. *Human Evolution.* Oxford: Oxford University Press, 2005.

Wood, Bernard, and Mark Collard. "The Human Genus." *Science* 284 (April 2, 1999) 65–71.

Bibliography

Woodmorappe, John. "The Geologic Column: Does It Exist?" *Creation Ex Nihilo Technical Journal* 13 (1999) 77–82.

Wopfner, Helmut. "The Malagasy Rift, a chasm in the Tethyan margin of Gondwana." *Journal of Southeast Asian Earth Sciences* 9 (May 1994) 451–81.

Wyngaarden, Martin J. "Some of the Problems of Chronology in Genesis." *Journal of the American Scientific Affiliation* 7 (September 1955) 43–46.

Xu, Xing, Hailu You, Kai Du, et al. "An *Archaeopteryx*-like theropod from China and the origin of Avialae." *Nature* 475 (July 28, 2011) 465–70.

Young, Davis A. and Ralph F. Stearley. *The Bible, Rocks and Time*. Downer's Grove, IL: InterVarsity, 2008.

———. *Creation and the Flood*. Grand Rapids: Baker, 1977.

www.ingramcontent.com/pod-product-compliance
Lightning Source LLC
Chambersburg PA
CBHW051937160426
43198CB00013B/2196